WRITING THE RURAL: FIVE CULTURAL GEOGRAPHIES

Paul Cloke, Marcus Doel,
David Matless, Martin Phillips,
Nigel Thrift

Paul Chapman
Publishing Ltd

Paul Chapman Publishing Ltd
144 Liverpool Road
London
N1 1LA

British Library Cataloguing in Publication Data

Cloke, Paul J.
Writing the Rural: Five Cultural
Geographies
I. Title
304.2
ISBN: 978-1-85396-197-7

Contents

Preface

This book arises out of an ESRC research project devoted to an examination of the economic, social and cultural impacts of the 'service class' on rural areas. The project was an attempt to document these impacts through close empirical work in a set of three rural communities. But something happened on the way. We found that the 'rural' became a real sticking point. Respondents used it in different ways – as a bludgeon, as a badge, as a barometer – to signify many different things – security, identity, community, domesticity, gender, sexuality, ethnicity – nearly always by drawing on many different sources – the media, the landscape, friends and kin, animals. It became abundantly clear that the 'rural', whatever chameleon form it took, was a prime and deeply felt determinant of the actions of many respondents. Yet it was also clear that we possessed no one theoretical framework that would allow us to negotiate the 'rural' and articulate its diverse nature as a category. Indeed, we discovered important differences in our preconceptions about, and attitudes towards, rurality and rural change which could not be easily submerged in one overriding narrative of what was going on in the villages we were studying.

Looking back, the emergence of these differences and the significance we attached to them is less surprising than we found it at the time. Then, we were trapped by an orthodox acceptance of the conduct of research as consisting of a set of seemingly democratic discussions between different researchers engaged on a common project. These discussions would ultimately lead to an agreed narrative. This naïve view shrouds both the power relations inherent in 'research teams' and the powerful prevailing orthodoxy about the kinds of 'results' which research should produce. But one of the many legacies of the turn towards postmodern and poststructuralist philosophies and practices in social science is the greater acceptance of pluralistic narratives. Certainly if the people living in the villages we studied read the rural in myriad different ways, why should it be any surprise that we, as apparently detached researchers, should read the rural any differently? It is against this backcloth that each of the extended essays in this book has to be understood. Each essay is an attempt by the author to draw out the 'rural' by drawing on different traditions in social an cultural theory.

Introduction

Refiguring the 'Rural'

Paul Cloke and Nigel Thrift

The 'rural' has often come under scrutiny as a cultural formation. But recently the force of that scrutiny seems to have become particularly intense. There are, no doubt, numerous reasons why this is so. But three of these reasons seem particularly important. First, rural areas have undergone rapid change. Connected processes like agricultural restructuring, new environmental policies, the incursion of middle-class incomers, the commodification of the countryside represented by the rise of the heritage industry, and the accumulation of new media images, have all pushed at the limits of what can be defined as 'rural'. Second, the rural is ever more in evidence as a meditative resource. The flood of publications that canonise the rural – in the press, in local history society pamphlets, on television programmes – shows no sign of abating. Finally, the academic world has done its part. Since the publication of Raymond Williams, *The Country and the City* in 1975 a regular series of book and papers has appeared and have pondered on the meaning of the 'rural'.

The net effects of this renewed scrutiny have been clear. Most particularly, the meanings and intensities of the rural have multiplied. For example, the rural is now routinely linked to gender, sexuality and ethnicity in ways which would have been considered remarkable only a few years ago. In turn, as the differences have multiplied so the idea of the rural as a fixed location has faded. It is now an infinitely more mobile and malleable term.

That this is the case can be demonstrated by considering the history of the 'the rural' in the rural studies literature. It is only in recent years that the rural studies literature has begun to reflect on the differences of meaning and construct associated with the rural. Although it is not our intention to provide a detailed account of changing notions of the rural in academic discourses, it is nevertheless useful as a backcloth to the essays in this book to recognise briefly that the 'the rural' has been the butt of a series of different perceptions, social constructions and representations from academic researchers. This series of changes has been described as occurring in four principal 'phases' (see Cloke and Goodwin, 1993) each of which, thus far, has been shorter and sharper than the last.

The first phase equated rurality with particular spaces and functions. Here 'rural' areas were thought to be clearly identifiable through negative characteristics – that is, rurality was taken to be symonymous with anything that was non-urban – and positive characteristics, where important elements of rural identity were specified. Definitions of the rural during this first phase typically suggested three principal criteria:

(i) rural areas are dominated by extensive land uses (agriculture, forestry) or large open spaces of underdeveloped land,
(ii) rural areas have small settlements which are strongly related to the surrounding landscape and which are recognised as different from the urban by their residents,
(iii) rural areas engender a way of life which is characterised by a cohesive identity based on respect for the environmental and behavioural qualities of living as part of an extensive landscape.

Clearly, these functional ideas of rurality and rural space still retain some currency.

The second phase in the changing construction of 'the rural' was to replace these functional definitions with more pragmatic concepts suggested by the use of political economic approaches. The notion of functional areas was seen to be undermined as changes in these areas were increasingly linked into the dynamics of the national and international economy, with accompanying assumptions that economic and socio-cultural activities were being organised on a relatively aspatial basis. Seen from this perspective, the 'causes' of rural change were usually located outside of the apparently functional rural areas themselves, and the 'localities' debate appeared to confirm that although certain places achieved a uniqueness derived from local society within broader processes of restructuring, rural places on the whole did not represent distinct localities (see Newby, 1986). As a result, some researchers called for a doing-away with the rural as any kind of analytical category (Hoggart, 1990) while others retained it as a pragmatic investigative unit.

These two phases were clearly riding on the parallel tracks of the changing nature of 'rural spaces' and changing fashions in theory and research in the social sciences. This parallelism is clearly identifiable in a third phase in the academic construction of the rural, heralded by Mormont (1990). His interpretation of the changing relationship between space and society in relation to the countryside leads him to the conclusion that it is no longer possible to conceive of a single rural space. Rather, he suggests that there are a number of different social spaces which overlap the same geographical space and, accordingly, that rurality should be seen as a social construct, reflecting a world of social, moral and cultural values. Here, then, we can recognise a shift from previous approaches to the rural, which had celebrated the modern, to academic constructions of the rural which fall into line with the postmodern turn in social science. To accept the rural as a social and cultural construct allows the rural to be rescued as an important research category, as the way in which the meanings of rurality are constructed, negotiated and experienced will interconnect with the agencies and structures being played out in the space concerned.

The fourth (and current) phase in locating the 'rural' in wider streams of social science thinking has been through a poststructuralist deconstruction of different rural texts. For example, Halfacree (1993) suggests that the emphasis on social constructions of rurality is prompting academic discourses on the rural to be routed increasingly through lay discourses, allowing the voices of 'ordinary' people to be heard. Drawing on the work of Baudrillard, he argues that the sign (= rurality) is becoming increasingly detached from the signification (= meanings of rurality) and in turn that sign and signification are both becoming divorced from their referent (= the rural locale). If a characteristic of so-called postmodern times is that symbols are becoming ever more detached from their referential moorings, then academic interpretations of the rural should reflect that move as socially constructed rural space becomes increasingly detached from geographically functional rural space.

Our own work reflects part of the wider move in the rural studies literature towards consideration of the social and cultural construction of the rural. It sits alongside other discussions (see, for example, the debate in the *Journal of Rural Studies* between Philo (1992, 1993) and Murdoch and Pratt (1993, 1994) which are seeking to open up questions such as which 'other' rural geographies have been neglected and what the implications are of adopting postmodern and/or poststructural approaches to rural studies. Thus, the essays in this book are predicated on an increasing awareness of the complexities, ambiguities and ambivalences of the rural, and they seek to make a modest contribution to these discussions by addressing different facets of this increasing awareness.

David Matless' essay is chiefly sited in the growing historical literature that identifies the rural as a discourse concerned with problems of identity and, most specifically, the problem of imagining 'Englishness' (e.g. Colls and Dodd, 1986; Porter, 1992). Matless is intent on showing how a Foucauldian approach to this discourse can rescue the English village – one icon of Englishness – from being consigned to the dustbin of ideology and can, at the same time, set it in motion, showing that it consists of complex multiple 'echoes' rather than a fixed and immutable category.

Martin Phillips' essay is concerned with Habermas rather than Foucault and therefore takes a markedly different track. In particular, he outlines some aspects of Habermas' work and places particular emphasis on exploring the implications of his concepts of 'public' and 'private' sphere and his arguments over language, knowledge and power. He uses reformulations of these concepts and arguments to write new interpretations of the rural and seeks to preserve a critical impulse in rural studies when doing so.

Marcus Doel's essay is, like the essay by Cloke that follows it, violently self-conscious. It is an attempt to use Derridean traces to uncover what takes place at the limit of ideas of the world as text. In attempting to problematise representation, Doel is concerned to show how deconstructive writing can constantly withold the meaning of the rural, turning it into a heterotextual assemblage that never stops (Doel, 1993).

Paul Cloke's essay is part of the self-relective turn in contemporary cultural studies. Writers like Steedman (1986) and Probyn (1992) have pondered on the degree and the ways in which the self can be introduced into writing. Cloke uses his own self-accounting as a way of interrogating what he has written about the rural, and seeks to understand the prompts to his own geographical imagination which lead him to be concerned to retain elements of the structuring of opportunities alongside elements of the experiencing of lifestyles in his accounts of how political economy can be (en)cultured.

Finally, Nigel Thrift's chapter draws on three main literatures. The first of these is the work going on in social and cultural theory around boundaries – between humans, between humans and animals and between humans and machines – work which simultaneously problematises gender, nature and technology and which strives towards a view of the work as a mixed, joint experience in which our capacity to live with difference can be enhanced (Hall, 1993). The second literature is actor – network theory. This is a recursive sociology of process that has become strongly concerned with the question of what counts as an actor, and how agency is achieved (Law, 1994). The final literature is poststructuralism and, specifically, its attempt to privilege the middle. The intent of the chapter is to show that these three literatures simultaneously inform and are the result of an historical accretion of different kinds of socio-technology. In turn, it is possible not only to problematise the rural as a fixed and stable construct but also to begin to show how the location of new and human communities founded in what would once have been conceived as monsters can become the normal currency of everyday life (Sheehan and Sosna, 1991).

As has already been noted, the rural is changing out of all recognition. Its economy is restructuring. Its society is on the move. Its culture is increasingly im-mediated. But the essays in this book are not just attempts to redescribe these changes. They are also attempts to understand them as part of a wider cultural problem: the location of the 'rural' as the subject and object of new meanings, conventions, and strategies of dissemination. As essays they are hardly exhaustive of this project, of course. There is too little in the essays on race and ethnicity, on gender and sexuality, and on registers of sense other than the visual. However, we hope that the essays at least begin to address new kinds of 'rural geographies' on which others can build.

REFERENCES

Classen, C. (1993) *Worlds of Sense. Exploring the Senses, in History and Across Cultures*, Routledge, London.

Cloke, P. and Goodwin, M. (1993) Rural change: structured coherence or unstructured incoherence? *Terra*, no. 105, pp. 166–74.
Colls, R. and Dodd, P. (1986) *Englishness. Politics and Culture 1880–1920*, Croom Helm, Beckenham, Kent.
Doel, M. (1993) Proverbs for paranoids: writing geography on hollowed ground, *Transactions of the Institute of British Geographers*, Vol. 18, pp. 377–94.
Gruffudd, P., Anthropology and Agriculture: Rural Planning in Wales between the Wars, in M. Heffernan and P. Gruffudd (eds.) A Land Fit for Heroes: Essays in the Human Geography of Inter-War Britain, Loughborough University, Department of Geography, Occasional Paper. no. 14, pp. 80–109.
Halfacree, K. (1993) Locality and social representation: space discourse and alternative definitions of the rural, *Journal of Rural Studies*, Vol. 9, pp. 1–15.
Hall, S. (1993) Culture, community, nation, *Cultural Studies*, Vol. 7, pp. 349–63.
Hoggart, K. (1990) Let's do away with rural, *Journal of Rural Studies*, Vol. 6, pp. 245–57.
Law, J. (1994) *Organising Modernity*, Blackwell, Oxford.
Mormont, M. (1990) Who is rural? or, How to be rural: towards a sociology of the rural, in T. Marsden, P. Lowe and S. Whatmore (eds.) *Rural Restructuring*, David Fulton, London.
Murdoch, J. and Pratt, A. (1993) Rural Studies: modernism, post-modernism and the post-rural, *Journal of Rural Studies*, no. 9, pp. 411–27.
Murdoch, J. and Pratt, A. (1994) Rural studies of power and the power of rural studies: a reply to Philo, *Journal of Rural Studies* no. 10 (forthcoming, March).
Newby, H. (1986) Locality and rurality: the restructuring of social relations, *Regional Studies*, Vol. 20, pp. 209–15.
Philo, C. (1992) Neglected rural geographies: a review *Journal of Rural Studies*, no. 8, pp. 193–207.
Philo, C. (1993) Postmodern rural geography? A reply to Murdoch and Pratt, *Journal of rural studies* no. 9, pp. 429–436.
Porter, R. (ed.) (1992) *England and Englishness*, Polity, Cambridge.
Probyn, E. (1992) *Sexing the Self. Gendered Positions in Cultural Studies*, Routledge, London.
Sheehan, J. J. and Sosna, M. (eds.) (1991) *The Boundaries of Humanity. Humans, Animals, Machines*, University of California Press, Berkeley.
Steedman, C. (1986) *Landscape for a Good Woman*, Virago, London.
Williams, R. (1975) *The Country and the City*, Paladin, London.

Chapter 1

Doing the English Village, 1945–90: An Essay in Imaginative Geography

David Matless

To make a start,
out of particulars
and make them general, rolling
up the sum, by defective means –
Sniffing the trees,
just another dog
among a lot of dogs . . .
William Carlos Williams, *Paterson*

1. INTRODUCTION

This essay considers, through a close reading of a number of texts and images concerned with village life and landscape, a culture of landscape and settlement making up the village in England since 1945. In their writings, their drawings, their photographs, their paintings the authors and artists presented here are all 'doing' the English village. The word 'doing' has been employed for a number of connotations; procedure, transformation, including violent transformation, habit, ritual, dance. Of course not only those written of in this essay 'do' the English village; that work is not confined to writers and artists. The English village can be regarded as a mythic figure, one dancing in English and other imaginations, a figure where people have located emotions, wishes, houses, anger and more, a site of values which are by no means tied to a rural location. This essay seeks to take the English village seriously as a cultural figure, and to describe some of it.

In this description the essay engages with a number of prevalent notions, worth spelling out to begin. The bulk of accounts dealing with 'myths' of the English village employ that term in a derogatory sense, denoting falsehood, illusion, deception. The mythic and imaginative English village is seen as an 'image' distorting or concealing a social or economic 'reality'. I have no wish to deny the partiality of the images considered in this essay; I do not though want to treat images as lower in a hierarchy of truth than some *thing* more 'real' (see Matless (1992) for a fuller account of the approach pursued here). Rather such images are here considered as speaking their own particular and partial truths; they *may* be seeking to deliberately evade other truths, but need not necessarily be interpreted as such. One might term this challenging of analytical hierarchies of truth a challenging of a *rhetoric of reality*. Writers such as Howard Newby (1980) and Brian McLaughlin (1983), for example, tend to criticise images of the village by contrasting them with, and judging them against, a 'truer' reality. Thus a 'rural idyll' is seen to conceal a truer reality of, say, poverty and lack of amenities. The writer's task here is to demystify, to unveil, to, in McLaughlin's phrase, lift 'the lid off the chocolate box'. To challenge this rhetoric of reality is not to deny rural poverty, nor to deny any possible relationship between imagined idylls and a lack of political engagement with such issues; it does though allow for a *more* sustained and critical, less dismissive engagement with the many and varied myths, nostalgias and moralities of the English village and country. It is perhaps worth pointing out here that the rhetoric of reality is by no means recent. Take for example Cecil Stewart's 1948 *The Village Surveyed*, a text which would certainly warrant more detailed consideration if this essay were aiming at a comprehensive coverage of village writings. Stewart's is an account of Sutton-At-Hone, in Kent. Describing his 'Approach' to the place, Stewart immediately declares his intention to wipe away the associations of the word 'village' in order to present a truer picture: 'The confirmed town-dweller will recite dreamily a long string of nostalgic associations – thatched cottages, old grey church, elms, lichen and ducks-on-a-pond. . . . The first thing to be done is to sweep away all these romantic notions' (Stewart, 1948, p. 17).

One might argue that Stewart's is a reasonable strategy given his purpose of social survey, and one could perhaps say the same about elements of Newby's work, but such a rhetoric of reality has effected, with regard to critical analysis, to consign the imagined English village to a dustbin marked error. Accounts are either dismissed as wholly sentimental, or are sifted for their useful elements of information and their distracting elements of emotion. One can also criticise this approach for effecting to deny any place for the imagination in that village designated as 'real'. It seems sometimes as though environmental imagination is a delusory faculty possessed only by the incoming middle classes, while other residents simply exist.

This essay, by contrast, critically embraces the many imagined realities of the English village – its sentiments, its fantasies, its dreams, even its sugar-sweet pond ducks – as things real, powerful, political and moral; things serious and of importance in the culture of the country. Its purpose in so doing is in part to establish a *complexity* in the discourse of the rural. Too often images of the country are lumped together under a single simple category – 'the rural idyll' – as if there were no difference there, as if the culture of the rural in England were one of a simple sentimental homogeneity. It would seem that the rural simplicity and harmony supposedly conveyed by such images is assumed also to characterise the discourse in which they operate. Besides emphasising the complexity of the village, the essay also aims, in writing of settlement, to *unsettle*. A counterpoint of the rhetoric of reality described above is of course the celebratory approach of those who seek to locate in the village an essential England, a timeless expression of community, a beautiful order. This essay gives such essentialist themes their due, but does not seek to bolster them. Rather, in considering the difference and complexity in 'the' English village it seeks to unsettle, to disturb what might be cherished notions. Not by vigorously and polemically debunking myth, but by leading it gently on to show itself in its variety, its power, its anger, its vitriol, its love. The imagined English village is as much part of the English 'heritage' as a hedge or a church or a duck or a sunset, but, as Michel Foucault once wrote in his essay 'Nietzsche, Genealogy, History' when discussing the 'search for descent':

> we should not be deceived into thinking that this heritage is an acquisition, a possession that grows and solidifies; rather, it is an unstable assemblage of faults, fissures and heterogenous layers that threaten the fragile inheritor from within or underneath. . . . The search for descent is not the erecting of foundations: on the contrary it disturbs what was previously considered immobile; it fragments what was thought unified; it shows the heterogenity of what was imagined consistent with itself.
> (Foucault, 1986, p. 82)

The current essay indeed follows the precepts of that form of history Foucault termed 'genealogy' (Dreyfus and Rabinow, 1982, pp. 104–17; Matless, 1991a). Elsewhere in 'Nietzsche, Genealogy, History' Foucault wrote that 'Genealogy does not resemble the evolution of a species and does not map the destiny of a people' (Foucault, 1986, p. 81). This essay too does not developmentally chart a myth, or project a future imagination from a past continuity. Its intent is to gently unsettle, at the same time conveying a rich complexity and variety in the imagined English village.

Partly, it should be noted, that richness and variety is conveyed here by writing of things other than villages; landscapes, land, people, cities, towns, railways, films, plants, maps, history, etc. Villages after all only make sense in relation to these and other things. This is not though a case of sketching a contextual background *from which* the foregrounded village takes its meaning. Rather the various things are held at the same level, some to be brought forward at some time, others at another. Nothing is necessarily foregrounded here. Another strategic point to note is that despite emphasising heterogeneity in the imagined English village, this essay still maintains, in its title and elsewhere, a singular 'English village'. This is not intended to suggest a consistent object of study, a solid referent for the writings and images under scrutiny here, rather 'the English village' is kept in this essay as a phrase, generally without quotation marks, to remind the reader of what is a very real figure in the imagination. This is not to build the myth up, but to keep it visible, to acknowledge its presence, to remind of its power. This essay seeks to unpack the English village, not to deny it, or tear it apart.

This unpacking is done through a series of close readings of particular books and images. The works considered are not necessarily 'key' texts, though some it is suggested were especially influential, and they have not been selected to provide a precisely even coverage of chronology or perspective. Each does though express a particular set of sensibilities regarding landscape and settlement. This is not then a comprehensive overview of the English village from 1945 to 1990; rather it is an initial genealogy of a part of English culture. It is also an exercise in a particular form of analysis and writing, and some points on the procedure and composition of the essay need making here. Firstly there are really six essays gathered. Each work or set of works is considered for the most part in isolation from the others. The six are written in together for the conclusion, but in the main body are held apart, and might be read on their own as well as together. This is not of course to say there is no connection between the essays; explicit connections and contrasts are made at times, and indeed one could read the set as arranged in an open constellation, each set up as distinct, but able to be connected to every other, as well as to things beyond. And doubtless empty spaces, black holes, might be found. Thus though the essays are ordered chronologically, with one exception, their narrative is not only sequential but spatial.

The individual essays are also held distinct to allow each book or set of books and images a space to show itself. In his study of the place of Tibet in the Western imagination, Peter Bishop writes of a distinction between 'telling' and 'showing':

> It would be relatively straightforward to *tell* the story of the Western encounter with Tibet, the experiences of the explorers, their discourses, the meaning of their fantasies. But this is not intended to be just a study *of* the imagination, but *in* the imagination. I therefore want to show as well as to tell.'
>
> (Bishop, 1990a, pp. 22–3)

Bishop also writes of his quotation of descriptive passages 'not just as illustrations for a conceptual argument but as part of the construction of the argument itself, woven into the text, integal to it' (Bishop, 1990a, p. 23). The same holds for this study of the English village imagination. Such an approach not least of course demands giving a serious attention to the aesthetic of the writings and images under scrutiny, an attention I have argued is absent from much work on the English village, and allowing that aesthetic to show itself. One must show how accounts convince, how they articulate their power, allowing them aesthetic amplification and amplitude. There is of course though an element of danger in this procedure. One might manage to convey power so effectively that it might once more convince when you don't intend it to. But to refuse to confront the aesthetic power of images is surely to write out their effects, and thereby to fail to comprehend the very thing one purports to consider. The intention here then is to let that power show, deliberately; to allow it a space, but not set it loose.

Three further points should be made in introducing this set of essays. Firstly the individual essays vary greatly in length, reflecting what it was felt necessary to write to convey that which the particular books and images were saying towards the purposes of this essay. Notes on the selection of each are included at the beginning of their essays. Secondly none of the material under consideration here is narrowly 'academic' in its content or audience. All the books and images here had an ostensibly 'popular' purpose (on academic studies of village communities see Harper, 1989). That said though they all purport to speak with an authority akin to that of the scholar. All suggest being based on careful study rather than speculative musing. As such, the boundary of the scholarly and the popular is blurred in this material, as indeed it is in much of the discourse of the English village. This theme of the forms of authority embodied in the various accounts is returned to at the conclusion of the essay. The third point to be made here is that in writing just now of the 'discourse' of the English village (another term used in a Foucauldian sense; see Philp, 1985, p. 69; Matless, 1991a), I do not equate discourse with 'text'. While the written word conjures much of this discourse, languages of sight, smell, taste and touch also operate at the heart. Maps, colours, faces, textures, building materials, landscapes, scents, odours abound. In the English culture of rural settlement, the power of all senses in articulating the country, in making it an object of desire, and in drawing people to it, whether for play or residence, should not be underestimated.

Before beginning the first of the six essays, I will signal a few recurrent themes. These are returned to in conclusion, but are worth indicating at the outset. They include morphologies of settlement, desires to settle, versions of home, knowledges of locality, views of landscape, values of environment, languages of authenticity, categorisations of people, articulations of power. All these run through the essay, as do themes of environmental consumption, class relations, definitions of the rural, and definitions of England. This last demands some comment. The countryside has often been presented in England as symbolic of national identity (Wright, 1985; Howkins, 1986; Matless, 1990b), with the village as one of its iconic foci. Thus much of this essay touches on conceptions of Englishness. This whole essay is itself though also very particularly English, and its arguments cannot necessarily be extended to other places, though doubtless similar themes might be found elsewhere. To make comparative study is, however, beyond the scope of this essay. This particularity does not though strike me as a problem. If we are to take seriously themes of the partiality of knowledge, of the local formation and operation of narratives, of the geography of the imagination, then speciality in our own produce is to be practised rather than resisted; not that we might speak a private language, but a particular one.

2. SHARP'S *ANATOMY*

Thomas Sharp's *The Anatomy of the Village* appeared in 1946, continuing the pre-war visionary and modernistic strain of English landscape planning to the post-war English settlement. This was a particular kind of modernism, one which was simultaneously presented as preservationist, concerned less with replacing old with new but with maintaining old and building new in an orderly fashion. Order versus disorder, rather than old versus new, was the imaginative scheme of this planning and preservation (on this see Matless, 1990a, 1990b, 1991b; Potts, 1989). Sharp's *Anatomy* was an influential work, but it is considered here not just for its influence but to emphasise an important modernistic current in the imagined English village. The place of the English village in relation to ideas of tradition and modernity is addressed at various points in this essay, and at the conclusion; Sharp's village is included at the outset to stress the actuality and possibility of imagining the village in a modernism.

The *Anatomy* formed a rural complement to the several books Sharp produced on the replanning, and in some cases rebuilding, of various historic cities, including Durham, Oxford and Exeter (Stansfield, 1981, pp. 157–69). Sharp's book was initially prepared in conjunction with the wartime Scott Report on Land Utilisation in Rural Areas, but publication was refused by the Ministry of Works and Buildings, and it took eighteen months of pressure before it was produced, in a form now devoid of associations with the Ministry. The book subsequently sold 50,000 copies (Stansfield, 1981, p. 155), and helped lay a trail for a long genre of the geographical study of the form of rural settlement.

Sharp writes with a sure voice, individual rather than bureaucratic, in a certain language of modernist planning and design, delineating and recommending a certain form of settlement, mapping the future village 'anatomy'. The anatomical metaphor emphasises a functional unity in the individual village, contained and balanced in its own form, and operating in concert with similar settlements elsewhere, each (for Sharp's is a vision to apply throughout the country) in the image of the other, unique localities, yet linked in the principles of place anatomy. Sharp looked to a modern countryside, of planned, compact, nucleated villages; modern, yet at the same time grounded in a particular notion of tradition (on a 'modern' tradition see Matless, 1990a). And Sharp asserted that modern tradition (of which more below) against what he regarded as a contemporary malaise. Writing of 'The Village To-day', he presents the village as in social flux: 'The social structure of the village has changed; so much so indeed, that it is hardly any longer possible to speak of it as having a "structure", if that is taken to imply any quality of stability. The old sociological simplicity has gone' (Sharp, 1946, p. 24). Likewise, in design, formal simplicity had passed: 'the tradition of maintaining compactness in settlement . . . has to a considerable extent disappeared' (Sharp, 1946, p. 35). Sharp lamented scattered or ribbon development, unconscious, unplanned 'scrappily complex' growth, citing as worst of all seaside villages which no longer had 'a whole character'. In a pairing of maps of a Kent village (Figure 1.1) Sharp took the plan-view to contrast the results of 'Non-Planning' – 'bad' grouping, a loss of 'character and coherence', a road dividing the village – with what might have been; the settlement bypassed and good grouping maintained (Sharp, 1946, p. 52; on the plan-view in inter-war landscape planning see Matless, 1990b).

Sharp's riposte to the apparent muddle was a call for a 'conscious simplicity' in the planning of future settlement, conscious because 'simplicity' could no longer emerge without help from the planner; 'our new villages and our rebuilt old villages cannot in the future have the artless and unsophisticated simplicity of the natural growing villages of the past' (Sharp, 1946, p. 37). This conscious simplicity would for Sharp be a deliberate modern extension of a long 'English Tradition' of village form and fabric. Opening his *Anatomy* with a description of 'The Place of Precedent', Sharp presented the ideal type of the English village: 'Informal, it is nevertheless orderly. Utilitarian, it often possesses a remarkable beauty; or, if it does not have that, it generally has at least a charm and a pleasantness and a whole character' (Sharp, 1946, p. 5). Though considering it 'dull and improvident, unworthy of our heritage' not to 'conserve . . . traditional harmonies', Sharp was at pains to distinguish his progressive citation of a 'Place of Precedent' from an 'indulgent' and nostalgic approach to the English countryside:

> Respect for tradition is an excellent thing, provided that the tradition respected is a genuine living tradition. A true tradition is subject to growth and development. It is not a pool which has welled-up at some particular moment of time, and has remained stagnant ever since. It is a flowing eddying stream that is continually refreshed by new tributaries, a stream whose direction is subject to change by new currents created by new conditions.
>
> The tradition that is invoked to restrict activity in the countryside to the kind of activity which was common in the past is a false tradition. Any suggestion that new villages and extensions to existing villages should exactly follow the forms which often gave our old villages such beauty . . . Any suggestion that new village building should imitate that old kind of building . . . any hope to achieve, by planning, the exact effects which have resulted solely from a lack of planning; these would not only illustrate a sense of tradition gone morbid, they would also be doomed to failure from the beginning. . . .
>
> We must work out new forms to meet new needs and to use new possibilities.
>
> (Sharp, 1946, pp. 5–6)

The modernity of Sharp's 'flowing eddying' true tradition was emphasised in the designs displayed for new building. Sharp proposed using new materials and construction methods in 'harmony' with, though not in imitation of, existing building styles. Thus Sharp proposed as 'in character' a 'Modern Street' (Figure 1.2) whose style would subsequently be presented as quite alien to the 'English village'; a terrace of small brick houses, flat-roofed, simple,

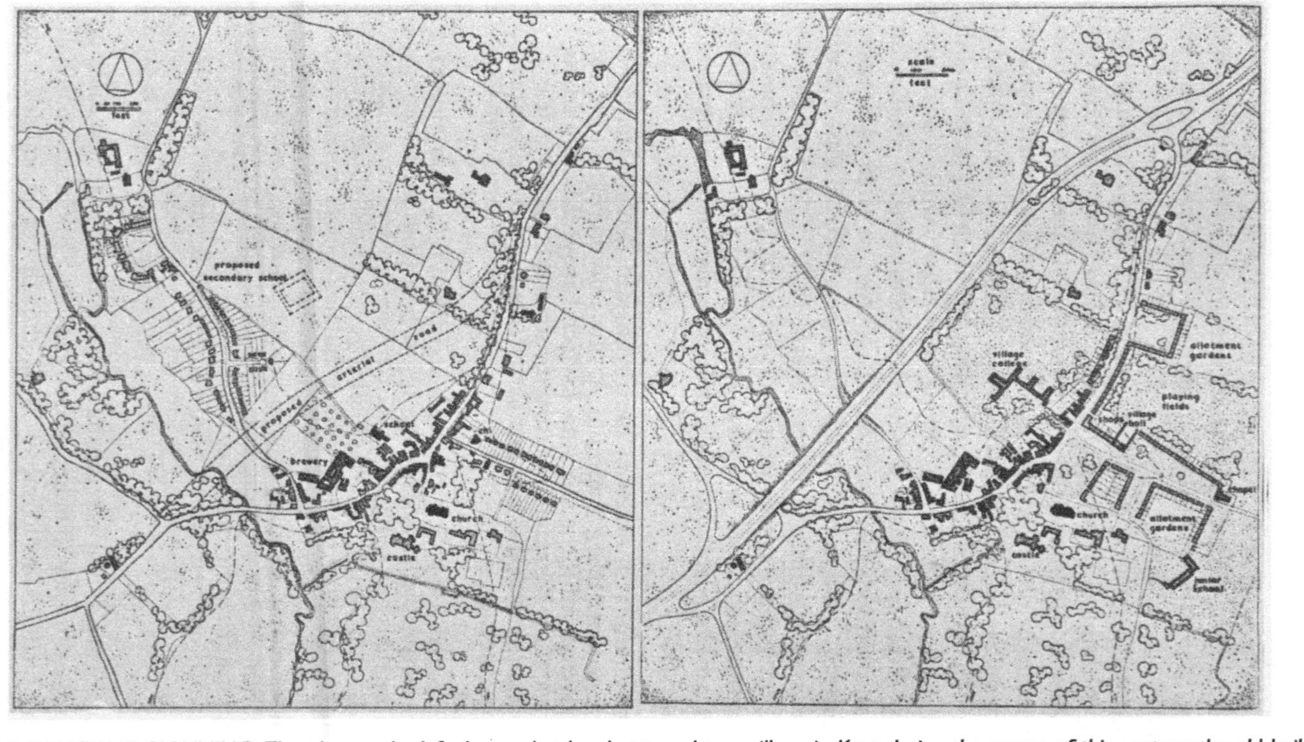

NON-PLANNING AND PLANNING. The plan on the left shows what has happened to a village in Kent during the course of this century: the old buildings are hatched: the modern buildings are in outline. Every possible mistake has been or is about to be made. The new houses are badly grouped, so that the village has lost character and coherence: an arterial is to cut the village off from its new school. The plan on the right shows what might have been done if the same new work had been planned: the village is kept compact: the new building shapes are in character: the by-pass is right away from the village.

Figure 1.1 'Non-Planning and Planning', from Thomas Sharp, *The Anatomy of the Village*, 1946.

Figure 1.2 'A modern street of small houses such as would fit well in character in a new or an old village', houses designed by F. R. S. and F. L. B. Yorke, from Thomas Sharp, *The Anatomy of the Village*, 1946.

bare, unadorned by gardens or cultivation, not at all picturesque cottage homes of England (Sharp, 1946, p. 51).

As indicated above though, the chief emphasis of Sharp's *Anatomy* was on layout, the book being replete with plan diagrams communicating village compactness. Sharp grounded his argument in what he termed 'The Psychology of Plan-Shapes', design principles which would produce 'the genuine village' and 'an orderly relationship of the parts' (Sharp, 1946, p. 63), with both parts and whole simple in character, and the plan-form 'immediately apprehensible'. A quality of definition and apprehensibility was for Sharp that which distinguished the village form from that of the town: 'They are not fussy, not elaborate, not complex and certainly not pretentious or monumental. A town plan is almost bound to have some or all of these attributes' (Sharp, 1946, p. 28). The village parts, considered Sharp, should be arranged in a pattern of enclosure, closing off views in and out of the village, and thereby contrasting with the wide-ranging views in the surrounding open country to provide 'a kind of psychological refuge and a visual satisfaction by way of contrast'. And the closure itself could be arranged so as to heighten (at least for the traveller, and the visiting planner) the sense of local 'climax': 'if that inward view is terminated on one of the public buildings, then the sense of climax will be heightened, and the traveller cannot but aware that he is entering a well established community' (Sharp, 1946, p. 65).

Sharp then put forward, in 1946, a planned visual psychology of English settlement, of grouping, climax, harmony and order; nucleated places of modern simplicity, minimally decorated, clearly defined and never 'fussy'. And Sharp consciously allied this minimal aesthetic to a particular morality of settlement, seeking to foster a particular orderly sense of community, a social as well as a visual 'orderly relationship of the parts'. In his *Anatomy*, building a modern English country on its 'Place of Precedent', Sharp laid bare to his many readers 'the patterns of grouping, the plan-forms, that are necessary or desirable for the collective functioning of the village and for its orderly appearance' (Sharp, 1946, p. 54). These themes of closure and grouping in the village would be echoed by later writers, but, as we shall see, the closed village would emerge as anti-modern, a psychological refuge from the modern world rather than a part of its network.

3. BONHAM-CARTER'S *ENGLISH VILLAGE*

Victor Bonham-Carter's *The English Village* was published as a Pelican in 1952. Bonham-Carter presents an 'official' version of the English village, though that version is of course as imaginative as any other. The book's back cover set up the author as just such an official voice, telling of an education at Winchester and Cambridge, a farm and an experience of local government, and displaying Bonham-Carter in angular and thoughtful profile, thinking on a pipe.

Much of what has been said above of Sharp's village also applies to Bonham-Carter's, indeed in *The English Village* the latter quotes approvingly from Sharp's *The Anatomy of the Village* regarding plan-forms (Bonham-Carter, 1952, pp. 100–1). As Bonham-Carter's title suggests, the 'English Village' is again treated as a type, a category which though not denying local variation does insist on encompassing diversity within one embracing generality. *An* 'English Village' sits within *an* 'English countryside', a national topic for the book's national audience. To read Bonham-Carter's Pelican meant engaging with settlement through a national vocabulary, the vocabulary of national planning. Planning, very much the agenda of the volume, was presented as an overdue post-war salvation for problems mounting long before: 'It seemed, in 1939, that the force of country life, as represented by the community of the village, was spent' (Bonham-Carter, 1952, p. 92). Bonham-Carter's planning, in contrast to Sharp's, prided in classification. Five 'Elements' constituted 'the village': the Physical, the Industrial, the Administrative, the Religious and the Human. Bonham-Carter identified in the latter definite temperamental and philosophical differences between the 'countryman' and the 'townsman', differences of dialect, and differences of physical type. Bonham-Carter employed the themes of physical anthropology developed by such writers as H. J. Fleure (see Gruffudd, 1988) and Harold Peake (the author of an earlier *The English Village* in 1924), calling up height, hair colour and head shape to answer the question: 'is there a distinct type of rural man?' The author thought that while one could argue that 'certain physical and mental peculiarities of the early races that inhabited Britain were transmitted to their successors, and are still recognizable about the country today', as far as 'ethnics' were concerned, there was 'no strictly identifiable racial type' (Bonham-Carter, 1952, pp. 201–3).

Two important differences should be emphasised between Bonham-Carter and Sharp, differences in procedure and in the conception of the aesthetic. Both jettison key elements of Sharp's particular modernism. Bonham-Carter's is a language more technocratic and less visionary than that of Sharp. Sharp had emerged in the period preceding the consolidation of a planning profession, where planners acted as self-consciously individual consultants, stamping a personal imprint large on programmes and designs. Bonham-Carter by contrast hailed from the more bureaucratic field of local government; his is a vocabulary of classification and segmentation, committee and control, contrasting to Sharp's unified vision and its slicing of red tape by the individual planner. And allied to this procedural difference is a difference in the conception of the aesthetic. Though like Sharp regarding a village's shape as having 'a direct bearing upon its vitality', particularly in relation to the 'creation of a strong focus' (Bonham-Carter, 1952, p. 233), in relation to other 'elements' of village life Bonham-Carter holds function and aesthetics apart, a separation not present in Sharp's *Anatomy*. Of agriculture for instance Bonham-Carter writes of how 'farming must, like any other industry, be economically sound. It must not and cannot be preserved for the sake, say, of tradition or aesthetics. The country is not a museum' (Bonham-Carter, 1952, p. 228; unlike Sharp, Bonham-Carter here also allies tradition and the museum). What happens here is less beauty being defined as necessarily against utility than function being defined as not necessarily having any relationship to beauty. Utility loses the necessary aesthetic component that it has in Sharp's work and in much of the pre-war town and country planning movement, where a main criterion of beauty was indeed that of 'fitness for purpose'. Function and form, economy and aesthetics, are detached, and with them fracture progress and preservation, the 'modern' and the 'traditional'. In such a fracture, of which Bonham-Carter was certainly in no way the 'originator' but rather a discursive symptom on his particular field, would operate much of the commentary on English landscape and rural settlement in the subsequent forty years.

4. LUCCOMBE OBSERVED

We now consider an account showing quite a different English village. *Exmoor Village*, published in 1947, was 'A General Account by W. J. Turner Based On Factual Information From Mass-Observation' of the village of Luccombe. *Exmoor Village* conveys an anthropological approach to English rural settlement, in keeping with the origins and intentions of Mass-Observation. That body had been established in 1937 when the anthropologist and ornithologist Tom Harrisson, co-founder with Charles Madge, resolved, on returning from the cannibal isles of the South Pacific, to explore the beliefs and habits of the people of Britain (Matless, 1989; Chaney and Pickering, 1986; Thrift, 1986; Summerfield, 1985). Mass-Observation utilised paid investigators and voluntary observers to record British society in a professedly scientific manner; *Exmoor Village* provides the only published Mass-Observed documentary of an English village. Published by George Harrap in a series entitled 'British Ways of Life', *Exmoor Village* presents the life of Luccombe as a type of society distinct from other types of British society, with the detail of its distinction to be deciphered only through a period of residence, through an 'in-depth' rather than a 'superficial' survey. Turner wrote of 'Our Method' as one of observers living 'for some time on the spot, inconspicuously and in varying roles. Their practice is not to resort to formal interviewing, but to gather information in a slow, roundabout way which gives a truer insight' (Turner, 1947, p. 12). 'Our Method' was one of the absorption of a place and its people – as much Mass-Absorption as Mass-Observation. This was a place and way of life defined by very distinct contours, by a local texture, and by its metaphorical and literal air, and all were to be absorbed. Turner's 'A First Impression' related how 'Walking down the street, savouring the air and light, one realizes that, however much photographs show, they (like our maps) never show enough, and they cannot convey the coolness and freshness of the air, nor the gentle declivity of the street' (Turner, 1947, p. 17). This is a place, and a method, which defies complete mapping, and which heightens senses other than the visual. It wants walking and smelling.

Luccombe was thus presented as an *other* way of life, though one nevertheless connected to the rest of the country; Luccombe was isolated in its 'parochialism', its local patterns of marriage, its once-weekly visit from a Porlock policeman (Turner, 1947, p. 25), yet had evident joins to surrounding Somerset and the rest of England. *Exmoor Village* included in an Appendix a 'Village Distance Chart' (Figure 1.3) showing the nearness and farness and connection of Luccombe to transport, services and leisure facilities. Six miles is presented as a long way away. Few villagers owned a car, and the bus only came within two miles.

The otherness of Luccombe is accentuated by *Exmoor Village*'s presentation of its topographical setting. Luccombe is given as a village in a valley to be discovered, or stumbled upon, by the visitor coming over a surrounding hill. The written-up survey is itself framed by the initial discovery and final sudden

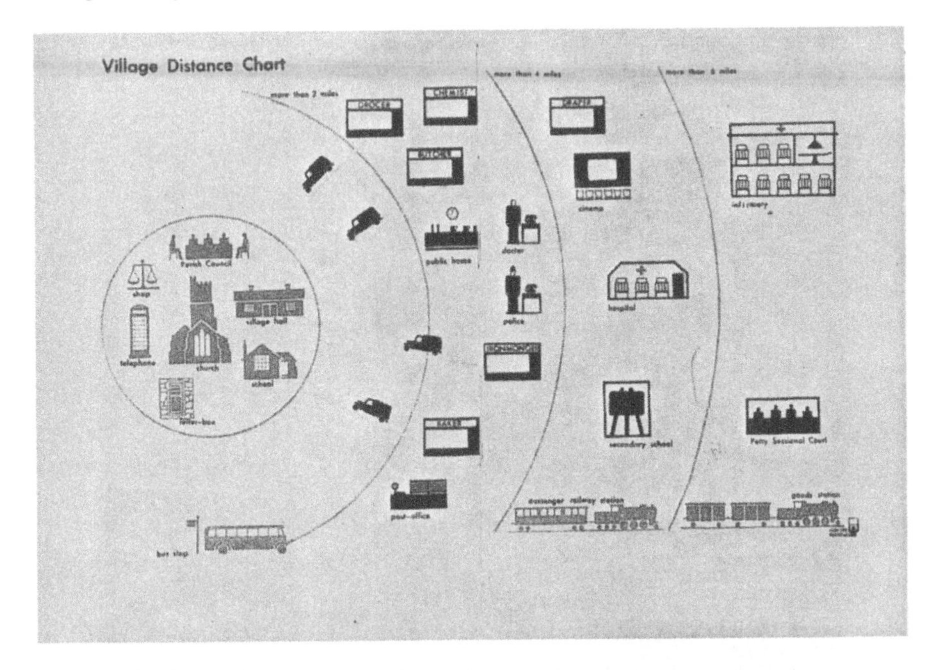

Figure 1.3 'Village Distance Chart', designed by The Isotype Institute, showing Luccombe's distance from various services, from W. J. Turner, *Exmoor Village*, 1947.

disappearance of Luccombe from view as the surveyor leaves the village. Photographs illustrate a nestling place (on similar topographical conventions in the inter-war period see Matless, 1990c). 'A First Impression' is garnered by driving up to an overlooking hill, in this case actually shortly after arrival in the village, to consider the prospect of ridges and valleys, after which the surveyor comes 'down off the moor' on a track which becomes a street and leads into Luccombe (Turner, 1947, pp. 16–17). We are subsequently told of Luccombe's enclosure such that at a panoramic point just out of the village: 'Even from here Luccombe has almost disappeared: you only have to go about a quarter of a mile or less from it in any direction and it disappears completely' (Turner, 1947, p. 20). Topography then makes this a very local place. The name 'Luccombe', we are later informed, means 'closed vale' (Turner, 1947, p. 95).

Such a 'closed' situation is portrayed by Turner as not special to Luccombe or Exmoor but as typical of 'The English Agricultural Village' (the title of the book's first, general chapter) as a whole. England, Turner tells, has 'only a few areas of flat country', in the Midlands, East Anglia and parts of Lancashire. Three-quarters of England, Turner estimates, is hill and valley country, where 'every few miles . . . the country changes in character, sometimes quite radically' (Turner, 1947, p. 13), and from such a topography is held to emerge a pattern of local difference, agricultural diversity and cultural conservatism.

Figure 1.4 '"The Square", showing the Lych Gate and the Keals' Cottage', photograph by John Hinde, from W. J. Turner, *Exmoor Village*, 1947.

Luccombe possesses then, in its closed vale, not an exceptional or eccentric individuality, but a typical distinction.

Related to the emphasis on local closure is an emphasis on recording the individual detail of Luccombe, in particular the detail of domesticity. Within *Exmoor Village* the detail of domestic life is foregrounded in both text and photograph. And this is a detail not only made up of the inanimate adornments of walls and rooms but of the people of Luccombe themselves, grouped at their

Figure 1.5 'A Village Interior', photograph by John Hinde, from W. J. Turner, *Exmoor Village*, 1947.

work and their leisure. Just as the village is gathered in the valley, so the villagers are shown grouped in the private and public spaces of the village. Non-family as much as family groupings are shown, gatherings of community signifying kinship in place as much as in blood. Village women pause to converse on doorsteps, the shopkeeper shows patience before deliberating children, village children form a ring in a 'square' whose size and irregularity belies its title but which still serves as a place for gathering (Figure 1.4).

And in 'A Village Interior' (Figure 1.5) we find mother and children at the hearth. John Hinde's photographs, very evidently composed, not at all stumbled upon, present tableaux of the everyday.

In keeping with Mass-Observation's desire to announce the insignificant, the detail of furnishing and possession is also documented. An Appendix lists at length the 'Contents of Two Typical Luccombe Bookcases'; another catalogues 'A Typical Luccombe Interior', taking us around furniture, floor, window-sill, mantelpiece, 'Below Mantelpiece, above Fire', the dresser shelves, 'Hanging on the Dresser-shelves', and the walls. We are told, for example, of chair coverings, one ash-tray, antlers and a plant-stand shaped like a black cat. Similar detail fills that part of the main text concerned with interiors (Turner, 1947, pp. 26–9). Tables of the types of wall-covering to be found in the village are compiled, and the collections of knick-knacks and brass ornaments distinguishing most cottages are noted. The external fabric of the built environment is also recorded in its particularity; differing styles and colours, the tone of moss on various walls. Houses in Luccombe are presented as individual, and a selection is reproduced in colour.

These details are not simply recorded 'for their own sake', though, but as part of a wider argument about rural difference. Throughout this description of interior and exterior Luccombe an emphasis is placed on the fact that neither conforms to standards. This is not the result of any architectural or design rebellion, merely that there are no pre-ordained standards to conform *to*. The exceptions are two council houses: 'the inevitable semi-detached rough-cast boxes . . . without the slightest trace of character or particular local appropriateness . . . not in the least offensive, but they are out of character' (Turner, 1947, p. 18). The non-standard is seen to produce character, and it is in this that Luccombe's 'character' is seen to lie. Such arguments concerning standards and distinction are not confined to matters architectural. Turner also argues against nationally standardised education being practised in local schools on the grounds of the fundamentally different needs of children in this 'way of life' distinct from the world of the town. Criticising 'the persistent bad influence of the nineteenth-century mania for urbanizing country people', Turner laments: 'It does not seem to occur to anyone that a farmer requires a different sort of education from a clerk and that it is the standardization of education that should be deplored. For there is a real and deep contentment with life in Luccombe' (Turner, 1947, p. 83). Turner's Luccombe *contains* country children born to a country life, almost pre-destined, and seemingly without the option of migration.

Produced in 1947, in the austere flush of post-war reconstruction, a time of social-democratic visions and plans for nationalisation, such an emphasis on local distinction and difference could itself stand out as distinct from the tone of the Scott Committee and the 1947 Town and Country Planning Act, from the planning vision of a Thomas Sharp, and from the general intent to plan a future of modern rural settlement. Indeed in many ways Turner presented Luccombe as standing for a way of life running counter to that of a modernising Britain. At times this presentation takes on a note of elegy, a tone of lament for a 'natural' way of life passing away. Turner's Luccombe here sits alongside the front bars of Bolton pubs fondly documented in Mass-Observation's 1943 *The Pub and the People* as threatened reservoirs of male, unAmerican working-class behaviour (Matless, 1989; Mass Observation, 1987). Turner wrote that in the agricultural village, motor transport and changes in farming had 'revolutionized the habits of centuries, dissolving local ties and uprooting families more thoroughly even than wars'. Such change was, he noted, often ultimately 'for the better', but led to a disruptive period of transition, with 'the ordinary way of life' 'in some respects' deteriorating. Thus there was 'less character', the 'communal life' had 'dwindled from what it was', and, with the decline in power of the manor and the rectory, there had been a loss of leadership: 'The village has . . . lost its aristocracy and become more of a proletariat, so the distinction between urban and country men is rapidly diminishing.' 'Ancient leaders' had gone, with nothing left in their place 'except the ravings or wisdom of the daily newspapers competing for the villagers' pennies'. Thus distant commercial concern had replaced church and manor's proximate 'natural functions' (Turner, 1947, pp. 14–15). Despite these two pages of lament early in the book, though, the pervasive tone of *Exmoor Village* is more one of Luccombe standing not for a thing past or slipping away but for a continuity, leading from the past through the present and into the future. Turner detects through perusing the parish records an 'astonishing stability of life in this part of the country. Indeed Luccombe must be considered as one of the villages in England which has changed least from the Norman Conquest in 1066 to the present day' (Turner, 1947, p. 90).

This stability he traces directly to the agricultural economy, the main occupations having remained 'almost entirely unindustrialized. The majority of the villagers of Luccombe were doing in 1945 very much what their forefathers were doing in 1145.' Inaccessibility, and a lack of resources to attract invaders, are also cited as factors in stability (Turner, 1947, p. 90). This extraordinary continuity Turner finds manifested in the parish church, the built complement to the parish documents. Turner acknowledges the poor contemporary attendance (Turner, 1947, p. 49), and sees the church less as standing for a spiritual than a local historical continuity. Turner writes in terms as much topographical and architectural as devotional when he states that 'The church is still the centre of the village as it was in the 14th century and earlier.' The church lends built expression to both continuity and locality, of which Turner writes: 'to get some sense of its age and the long continuity of its life, we had best enter the parish church, whose square tower marks its site conspicuously in the plain below the moors and Dunkery Beacon' (Turner, 1947, p. 86). In a chapter on 'The Surrounding Country and its Life', Turner, attending a service in Porlock, similarly bears witness less to God than to local England:

> Service at an old English parish church with a good rector is one of the most pleasing and moving experiences an Englishman, or anyone of

> British descent, can have. Here alone we can be at the very heart of our local history in its continuity over a period of six or seven centuries.
> (Turner, 1947, p. 97)

Luccombe then is invested with a permanence, given a stability contrasting with, and holding lessons for, the changing world around. Not an unchanging rock in a turbulent sea, but rather a way of life which might persist, maintaining the old and assimilating the new when required, after other ways of life had played, or burned, themselves out. Turner wrote of how readily 'completely scientific innovations can be assimilated into the old-established life of the countryside'. The motor bus, for example, had removed 'isolation' and increased mobility 'yet the people of Luccombe remain as definitely rural as ever' (Turner, 1947, p. 94). Here was permanence, memory, Time itself, and for Turner a certain consolation:

> In a world where so much changes so rapidly as almost to obliterate memory, it is good to think that Time has a foothold in Luccombe where the continuity of the life of individual men and of their families imparts a sense of permanence and duration to all things.
> (Turner, 1947, p. 103)

In all of this, Turner presents Luccombe as exemplary of a 'country' as opposed to, and defined in its contrast to, an 'urban' way of life. The opposition is presented explicitly as one of meaning versus the meaningless. Turner cited 'the general dissatisfaction' in England and Europe 'with what life has come to be in all its meaninglessness' as signifying 'a coming fundamental change', and introduced Luccombe, and its mass observation, as a beacon for the new future:

> some of the things which all men desire were once to be enjoyed more in our Exmoor village than they are anywhere in town or country to-day; and this description of things as they are in our village may throw some light on what is needed for the future.
> (Turner, 1947, p. 15)

The permanence of meaning resident in Luccombe was contrasted to what were regarded as the whims of fashion and politics, urban ephemera less enduring than the deeper reality of the country. Luccombe becomes a site of the real, a place of vital meaning rather than of the superficial language of the media and the politician. It should be noted here that it is those engaged in agriculture who are considered to form 'the real village community' (Turner, 1947, p. 17); not the parson, not the schoolmistress, and not the two old ladies who have retired to the village, all of whom are touched by outside and not of the land. This contrast of the authentic and organic culture of a way of life and the inauthentic manufactured voices of media and politics (the play of agricultural and industrial terminology is obvious here) echoes earlier Mass-Observation works, notably *Britain* of 1938 (Matless, 1989). In *Exmoor Village* the analysis is taken so far as to literally exclude politics from the 'real' village. In Luccombe:

> Life is . . . very much stripped of all its superfluities, and most of the questions that are hotly debated in cities and big industrial centres have no interest whatever for Luccombe people as they have more serious business of their own to attend to. There are strictly speaking, therefore, no political opinions or discussions.
>
> (Turner, 1947, pp. 30–1)

Socialism in particular is presented by Turner as alien to the place. Though politics and the media did not 'belong' in Luccombe, Turner nevertheless saw them, and urbanity in general, as posing an alien threat to the way of life. Turner observed that the 'political campaigns' of the newspapers meant 'very little' to 'the villager': 'they hardly seem to touch on his vital interests. He does not even know where his vital interests lie, for he is likely to find the "circuses" of the city as seductive in prospect as the townsman has found them devitalizing in effect' (Turner, 1947, p. 15).

The urban reaches out to devitalise, sapping, with its seductive power, physical and moral fibre, literally demoralising the 'countryman'. Turner though detected possible topographical rescue for the country, arguing, in relation to the difference-generating terrain of the hilly three-quarters of England, that

> At a time in the world's history when we are witnessing a quick and steady deterioration in so many directions, we may be inclined to be thankful that Nature has here put a check on the tendency to degradation which exists in mankind along with the desire to improve.
>
> (Turner, 1947, p. 14)

Central to *Exmoor Village*'s presentation of Luccombe was the use of photography. The photographer John Hinde lived in the village like the other observers. The book contains twenty-nine colour and twenty-two black-and-white photographs. Turner wrote of 'the addition of colour' being: 'not in order to make pretty pictures, or to relieve monotony, but to bring home to the reader in the most vivid manner possible the precise nature and appearance of the people and their surroundings' (Turner, 1947, p. 12). The extensive use of photography in *Exmoor Village* in some ways parallels the plethora of 'Countryside in Pictures' books issued at this time, and perhaps alluded to by Turner in his 'pretty pictures' comment. Such volumes (a notable series was that produced by Odhams, including 'The British Countryside in Pictures' and 'Romantic Britain') were dominated by the conventions of the Picturesque, and in displaying the rural everyday simultaneously succeeded in making it into cliché. The tableaux of the everyday in *Exmoor Village* are not entirely distinct from such images, but sit outside the genre by virtue of the nature and quality of photographic reproduction, and their juxtaposition to other forms of recording. Hinde's vivid photographs stand out in their rich colour in a time of austerity and paper rationing, a time commonly recalled in later years as synonymous with the colour grey. With this colour Luccombe comes over as almost a place enchanted, the village and the people possessed of hue, bright

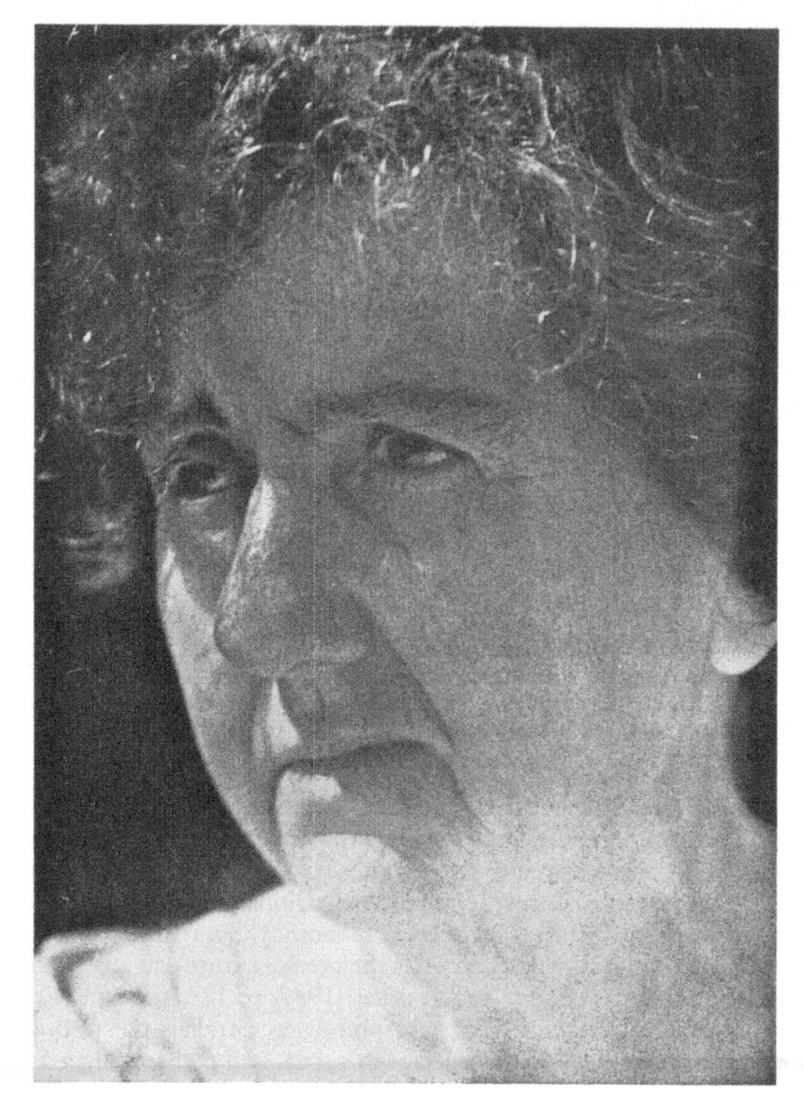

Figure 1.6 'Mrs Keal', photograph by John Hinde, from W. J. Turner, *Exmoor Village*, 1947.

with a natural make-up. Such a make-up shines especially in the several close-up photographs of individual faces (Figure 1.6). Living in Luccombe would seem to put colour in the cheeks. Such images might also be read by some at the time as documentary of head-shape and local physical type, but whether or not head dimensions are to be noted, these people glow from the pages of their other way of life.

These Luccombe people are nigh on superreal. As an organisation Mass-Observation had been since its inception linked to various currents of modernist thinking, including surrealism, in particular through the poet Charles Madge, co-founder of the body with Harrisson (Chaney and Pickering, 1986) Mass-Observation especially sought to experiment with the representation of everyday life, in the juxtaposition of a variety of 'artistic' and 'scientific' media, and in a broad intent to make strange what was close to home. James Clifford writes, in his essay recovering examples of 'ethnographic surrealism', of how 'Mass Observation envisaged a comprehensive ethnography of British popular culture conceived as a defamiliarized, exotic world' (Clifford, 1988, p. 143). Enchanted Luccombe sits in such a world, in both the display of its photographs and in the jostle of representational method. Though the intent to experiment is less evident in *Exmoor Village* than in some earlier Mass-Observation productions, and though the representational methods are less thrown together and more sorted out, there is still a variety – photographs, statistical tables, distance charts, furniture lists – which does not entirely dovetail, which can leave a tugging mixture of social science and art in the mind of the reader. This again sets Luccombe apart from the English countryside of, say, the Odhams series. There the entrance to the country is through the simple convention of the Picturesque; there is no such simple single way into Luccombe. Through *Exmoor Village* a number of different windows on to otherness crop up, each lending a different angle on the same set-apart world.

5. HOSKINS' LOCAL

We now consider W. G. Hoskins' *The Making of the English Landscape*, first published in 1955. Hoskins' work is notable in articulating an anti-modern English landscape, a particular and highly influential reading of history and the local, and an excavatory approach to the English village.

Hoskins presented *The Making* as a distinctly new way of seeing England: 'No other book exists to describe how the various landscapes of this country came to assume the shape and appearance they now have' (Hoskins, 1955, p. 13). Hoskins instead offered his 'new kind of history which it is hoped will appeal to those who like to travel intelligently, to get away from the guidebook show-pieces now and then, and to unearth the reason behind what they are looking at' (Hoskins, 1955, p. 14).

The Making then was presented as a 'popular' rather than strictly 'academic' text, at least in aiming to reach the 'intelligent' traveller. D. W. Meinig, records how *The Making* had little immediate impact. Not until 1970, with its reissue in a Pelican paperback, and with a subsequent television programme in 1973, did Hoskins' reading of the English landscape gain a wider audience (Meinig, 1979, pp. 195–202). Meinig's essay considers a number of Hoskins' publications (see also Matless, 1993); here the attention is narrowed to *The Making* alone. And while Meinig describes the delay in the book's 'impact' as 'the uncertain interval between seedtime and visible growth in a trial planting of a new kind of literature' (Meinig, 1979, p. 199), and quite reasonably confines his attention to a generic consideration of that type of literature, here Hoskins' work is considered less as an 'isolated' endeavour in 1955 than as a highly significant mark of an emergent English culture of landscape.

That Hoskins' method, of a combination of documentary research and of a fieldwork which he regarded not only as a source of knowledge but as an improving recreation, was entirely novel in 1955 is doubtful. Much the same set of themes can be traced in the 'progressive' regional survey movement of the 1920s and 1930s (Matless, 1990c). The novelty of Hoskins' work lies rather in his allying the pursuit of knowledge regarding landscape and place (indeed almost allying landscape *per se*) to a specifically anti-progressive and anti-modern outlook. Regarding landscape, Hoskins, unlike the earlier regional survey movement whose ideas informed much of the post-war planning system, presents beauty, and the local, in opposition to the modern. And the fracture of 'tradition' and the 'modern' identified as emerging in Bonham-Carter's *The English Village* gapes wide, though in a very different manner, in Hoskins' 'new kind of history'. Hoskins' local should also be distinguished from Turner's local Luccombe, a place also presented in opposition to elements of modern life. While Luccombe is anti-modern (and specifically anti-urban) by virtue of its particular exemplification of a country of continuity, in Hoskins' work the local appears as a concept anti-modern, anti-standard *per se*.

Hoskins makes his clearest opposition of beauty and modernity in the final chapter of *The Making*, on 'The Landscape Today' (Hoskins, 1955, pp. 231–5). While Hoskins is no admirer of the nineteenth century industrial scene (Hoskins, 1955, pp. 171–9), he saves his greatest wrath for the landscape change of the twentieth century: 'especially since the year 1914, every single change in the English landscape has either uglified it or destroyed its meaning, or both'. Only the reservoirs of the North and Midlands, he wrote, had 'added anything to the scene that one can contemplate without pain'. In his agony Hoskins intimates that he almost left his *Making* unfinished, but he brings himself to bear: 'It is a distasteful subject, but it must be faced for a few moments.'

What then for Hoskins constitutes the ugly and destructive, that opposed to landscape and beauty which in its opposition helps define the meaning of the latter as its contrary? There are various, associated villains, conspirators in modernity, in Hoskins' narrative of the 'landscape today'. Planning is one, whose role will be returned to below. Commerce is another, both in the person of the 'timber merchant' tearing the trees from a sold-off country estate, and, significantly, in the person of the farmer. Agriculture emerges in Hoskins as a possible (though not inevitable) enemy of the landscape. In such areas as East Anglia Hoskins deplores how 'the bulldozer rams at the old hedges', regrets the coming of a mechanical 'new ranch-farming' (an American allusion here) run by 'business-men farmers of five to ten thousand acres', with the yeoman ousted. Elsewhere Hoskins hoped that high hedgebanks and stone walls would be maintained in a more aesthetic 'economy' for the purpose of shelter and timber, and that 'the old field pattern' would thus remain. Alongside, though, and still more threatening than, the new agribusiness, a complex military–industrial alliance is presented conspiring to deface rural England. Politics, science and the military, whose alliance for Hoskins has culminated in the potentially apocalyptic destructiveness of a nuclear 'new age', are presented at odds with beauty, a beauty signalled for Hoskins by the landscape and the landscape art of England. Hoskins lists the violations of the Cotswold dip-slope, of Lincolnshire and of Suffolk by 'the villainous requirements of the new age': 'Over them drones, [in a sound without melody] day after day, the obscene shape of the atom-bomber, leaving a trail like a filthy slug upon Constable's and Gainsborough's sky'. Again there is American villainy in this landscape. This England of the new age, of the horrible present and frightening future, Hoskins catalogues as the

> England of the Nissen hut, the 'pre-fab', and the electric fence, of the high barbed wire around some unmentionable devilment. . . . England of the bombing-range wherever there was once silence. . . . England of battle-training areas . . . , tanks crashing through empty ruined Wiltshire villages; England of high explosive falling upon the prehistoric monuments of Dartmoor. Barbaric England of the scientists, the military men, and the politicians.
>
> (Hoskins, 1955, pp. 231–2)

Hoskins' is a story of past beauty destroyed by the present for a barbarous future. Past and present-future become antagonistic. One must save and cherish the former against the latter. Hoskins breaks any thought of a continuum of beauty, breaks with the progressive vision of Thomas Sharp and of the pre- and post-war planner-preservationists with their promotion of a modern tradition. While his vocabulary of defacement is strikingly similar to theirs, Hoskins holds out neither the actuality nor the prospect of a contemporary beauty. In *The Making*, tradition becomes confined in a past which is itself a refuge, a solace, to be cherished, its memory and its remaining artefacts to be preserved against the future: 'Barbaric England of the scientists, the military men, and the politicians: let us turn away and contemplate the past before all is lost to the vandals' (Hoskins, 1955, p. 232).

Hoskins' site of beauty and history is the local, and in this lies another of Hoskins' distinctive interventions in an English culture of settlement, in particular that of rural settlement. One can see Hoskins' preference for the local in, for example, his attitude to roads. Unlike the planner-preservationists of the inter-war years who admired its clean lines, its progress through the landscape (see Matless, 1991b), Hoskins detested the arterial road. One of his Englands of despair in 'The Landscape Today' was the 'England of the arterial by-pass, treeless and stinking of diesel oil, murderous with lorries' (Hoskins, 1955, p. 232). Of bypasses he wrote: 'They are entirely without beauty. Is there anything uglier in the whole landscape than an arterial by-pass road, except an airfield?' (Hoskins, 1955, p. 190). Hoskins preferred a lane, again unlike the inter-war preservationists who had seen the land and new road running side-by-side in complement. For Hoskins the lane, winding between fields, linking settlements, forms part of the fabric, the texture of the landscape. The new road, by contrast, is alien to the historic fabric, is not and will never be woven in. Hoskins had less objection here to the railway, indeed he admired the 'grandeur' of the nineteenth century works, comparing their scale of embankment and cutting to the hill forts of the Iron Age (Hoskins, 1955, pp. 198, 204). The railway though had acquired its greatest beauty through its subsequent dropping into the scenery, flowers overgrowing the marks of construction. Hoskins in particular evokes a branch-line beauty, describing, and picking out the smallest county in England for the setting, the 'pleasure of travelling through Rutland in a stopping-train on a fine summer morning':

> the barley fields shaking in the wind, the slow sedgy streams with their willows shading meditative cattle, the elegant limestone spires across the meadows, the early Victorian stations built of the sheep-grey Ketton stone and still unaltered, the warm brown roofs of the villages half buried in the trees, and the summer light flashing everywhere. True that the railway did not invent much of this beauty, but it gave us new vistas of it.
>
> (Hoskins, 1955, p. 206)

The branch-line here gives a vital passage to England, and indeed Hoskins' whole project in *The Making* could be characterised as a branch-line history. This is a metaphor not without historical allusion. *The Making*, along with Hoskins' 1959 manual *Local History in England*, culturally anticipates the surge in steam train and branch-line preservation from 1960 onwards, when the first closed line to be restored by a private society, the Bluebell Railway in Sussex, was reopened. Many other such private ventures followed. Like Hoskins' local history, steam and branch-line salvage find a virtue in not connecting with a wider network, in operating locally, back and forth along the same little line, and not beyond, in an atmosphere of rescue and reverence for something felt to have been passed by, bypassed, by the progress of the modern world. England of the branch-line, England of Trumptonshire, England of *The Titfield Thunderbolt*. A comparison of virtues and villains in that 1952 Ealing comedy and in Hoskins' 1955 *Making* is instructive. The film concerns the rescue of the oldest branch-line in the world from threatened closure by British Railways (Barr, 1977, pp. 159–65). The locals campaign to keep it open and run it themselves. A threat comes from a local bus operator who wants his service to enjoy a monopoly. Throughout the film the line seems to serve little 'economic' function. Its virtue is aesthetic and 'emotional'; these are its positive aspects, defined against that which it opposes. The oppositions are clear; state bureaucracy versus local community, 'human' value versus commercial value, local character versus national standards. Not that the story is anti-national England. Rather a particular definition of national England emerges at the end of the film from the preceding evocation of local England. The line is accepted as part of the network, and main-line drivers salute the local by tooting their whistles. Through the narrative of Titfield comes the hope of the redemption of England from a state of bureaucracy and commercial values, once more to be a place of localities.

One cannot trace a direct influence of *The Titfield Thunderbolt* on W. G. Hoskins, or indeed of Hoskins on the subsequent moves for branch-line preservation, but around all three one can detect a particular English culture of landscape and the local emerging. One can certainly though regard Hoskins' work in the 1950s and after as the catalyst for a 'discovery' of local history (Riden, 1983, pp. 21–2), for a popular upsurge in the local study and local recording of local event (for general thoughts on these themes see Samuel, 1989, pp. xlii–l). The metaphor of the branch-line, which I do not mean to be a solely critical one, can also I think serve for such history as well as for its mentor. In 1948 Hoskins helped found the Department of English Local History at Leicester University, was appointed Reader there, left in 1951 to become Reader in Economic History at Oxford University, and returned to Leicester to the Chair of English Local History in 1965. He retired from the post in 1968 (Meinig, 1979, pp. 198–9). In the late 1980s the Leicester Department sought to attract students to its (wide-ranging) M.A. course in 'English Local History' with a telling poster (Figure 1.7), an image which retains and sustains the ethos of Hoskins' *Making* of thirty years previous. A house is seen down a narrow lane, high-hedged, which turns from view, who knows where, before reaching the dwelling. English Local History sits in an old landscape down the end of a lane, seemingly away from any main highway.

Hoskins' local then sits, with its beauty and history, away from and against a version of the modern world. One can also read such aversion in Hoskins' telling of economic transactions. While he decries the scenic results and driving motives of commerce and industry, he warms to the local market embodying local transaction. Describing contemporary markets, with their stalls and shouts, Hoskins observes that 'all this is purely medieval' (Hoskins, 1955, p. 225). And the setting of the market, the English market-town, is also regarded as a local place of beauty and intriguing history (Hoskins, 1955, pp. 224–30). Hoskins ends his chapter on 'The Landscape of Towns' with a call for further investigation, leaving his reader, on the page before he launches into 'The Landscape Today', with a message of peaceful retreat:

> in the meantime how pleasant it is to find oneself arriving in the evening for the first time in some lovely little English market-town, where one can forget for a while the noisy onward march of science, and settle down to meditate upon the civilised past.
>
> (Hoskins, 1955, p. 230)

Against the modern world, Hoskins' local lies also in key opposition to planning. Again in contrast to the inter-war planner-preservationists, Thomas Sharp and the post-war planning system, the latter firmly in place by the time of *The Making*, Hoskins presents planning and beauty, planning and the historic, and planning and the local as opposed. Conveying 'The Landscape Today', Hoskins details the dissolution of a country estate. If such an estate happens to lie near a town, he posits, it is likely to fall prey to 'the political

Figure 1.7 'English Local History at Leicester', poster issued by Department of English Local History, University of Leicester, to advertise its M.A. and research courses, c. 1988.

planners', occupying villains who 'swarm into the house, turn it into a rabbit-warren of black-hatted officers of This and That'. Planning and bureaucracy, exercising the functions of the state, are set anonymously against the local scene. Local villagers, says Hoskins, 'detachedly' term the planners '"the atom men," something remote from the rest of us':

> And if the planners are really fortunate, they fill the house with their paper and their black hats, and their open-cast mining of coal or iron ore simultaneously finishes off the park. They can sit at their big desks and contemplate with an exquisite joy how everything is now being put to a good use. Demos and Science are the joint Emperors.
>
> (Hoskins, 1955, p. 231)

While Hoskins here seems to lament the dissolution of a house and its estate, in earlier chapters of *The Making* he paints a less than rosy picture of the creation of landscaped parks (Hoskins, 1955, pp. 126–37). While the earlier planner-preservationists, along with many other 'country' writers, had regarded the estates of the eighteenth century as a high point of English design, an historic climax of appearance (Matless, 1990a), Hoskins portrayed the parks as wasteful, and their creation as destructive of a local landscape, with villages and cornlands destroyed: 'Building themselves magnificent houses, they [the 'territorial aristocracy'] needed (or thought they needed) more square miles of conspicuous waste to set them off' (Hoskins, 1955, p. 133). Hoskins presented his own work as countering 'the wholly inadequate view that the English landscape is "the man-made creation of the 17th and 18th centuries"' (Hoskins, 1955, p. 135), and suggested his own high point of landscape making (though not landscape *planning*) as the period between 1570 and 1640, 'The Flowering of Rural England', the 'Great Rebuilding' (Hoskins, 1955, pp. 119–26). Hoskins lauded this time as a triumph of the vernacular, of local variety, of use and beauty combined, of 'minor' building in a peasant culture:

> In these two generations or so, the rich variety of regional styles of building – the vernacular of the English countryside – established itself everywhere, based upon the abundant local materials that a peasant economy, a peasant culture indeed, knew how to use well and beautifully. If we are to study and record the variety of minor English building before it is too late, in both country and town, it is in these generations that we shall find our richest evidence.
>
> (Hoskins, 1955, p. 122)

The vernacular, and its local richness, is presented as opposed to the standard and characterless landscapes seen by Hoskins as characteristic of the modern and the planned landscape, including the planned landscapes of the eighteenth century. Hoskins' narrative of English history supports his portrayal of planning as a force not for the enhancement but for the suppression of local difference.

What then constituted Hoskins' local vernacular richness? How did Hoskins characterise the local, apart from simply as the antithesis of the modern and the planned? In *The Making* the local sits above all as the site of detail, of depth and of meaning, opposed to the standard, depthless and meaningless 'Landscape of Today'. Walking on the border of Lincolnshire and Rutland, Hoskins points to the pleasure, intellectual and aesthetic, in detail, in the part in itself as much as in its relation to the whole: 'So, behind every generalization, there lies the infinite variety and beauty of the detail; and it is the detail that matters, that gives pleasure to the eye and to the mind, as we traverse, on foot and unhurried, the landscape of any part of England' (Hoskins, 1955, pp. 160–1).

Earlier in *The Making*, Hoskins employed the analogy of music to emphasise detail, though in this case the detail is held up for regard both in itself and as a means to understanding the larger England: 'One may liken the English landscape . . . to a symphony, which it is possible to enjoy . . . without being able to analyse it in detail. . . . The enjoyment may be real, but it is limited . . . and in the last resort vaguely diffused in emotion' (p. 19). Hoskins argued that if one could be more specific, could 'isolate the themes as they enter', unpacking the overall harmonies and perceiving 'the manifold subtle variations', then 'the total effect is immeasurably enhanced. So it is with the landscapes of the historic depth and physical variety that England shows almost everywhere.' And England, emphasised Hoskins, provided not only 'a programme of symphonies', 'magnificent views over a dozen counties', but the smaller and simpler 'chamber music of Bedfordshire or Rutland'. And there was 'perhaps . . . a more sophisticated pleasure in discovering the essence of these simpler and smaller landscapes' (Hoskins, 1955, p. 19).

The analogy of landscape as music begs the metaphor of reading. Hoskins presented England as a historical document to be read, and offered his means to the reader: 'The English landscape itself, to those who know how to read it aright, is the richest historical record we possess' (Hoskins, 1955, p. 14). And a document through which one could read this landscape was the Ordnance Survey map (on the role of the O.S. map in the 1920s and 1930s see Matless, 1990c), to be consulted either in the service of fieldwork or for its own sake. Hoskins' description of map-reading is worth quoting at length, embodying as it does his love for local 'flavour' and, in the sense of both taste and fragility, local delicacy. Hoskins' is a loving dissection of map and landscape:

> There are certain sheets of the one-inch Ordnance Survey maps which one can sit down and read like a book for an hour on end, with growing pleasure and imaginative excitement. One dwells upon the infinite variety of the place-names (and yet there is a characteristic flavour for each region of England), the delicate nerve-like complexity of roads and lanes, the siting of the villages and the hamlets, the romantic moated farmsteads in deep country, the churches standing alone in the fields, the patterns made by the contours or by the way the parish boundaries fit into one another. One dissects such a map mentally, piece by piece, and in doing so learns a good deal of local history, whether or not one knows the country itself.
>
> (Hoskins, 1955, p. 76)

Hoskins then describes in detail the 'exciting' and 'beautiful' map of the country around the Wash. There remains a whole book to be written on the roles of the map in English culture, a book in which Hoskins would certainly loom large.

Beneath (often literally) Hoskins' local detail is local depth. Hoskins' local is a deep resource, a place to be delved, a palimpsest to be peeled to show rich history. And depth is accentuated by settlement. The local is a settled rather than a dynamic site, which has accrued over the years, and which sits still for scrutiny. It is not a slippery customer. Hoskins employs an organic metaphor for a place's history, writing of a 'cultural humus' (Hoskins, 1955, p. 235).

Depth, detail and the organic all combine in the final pages of *The Making*, where Hoskins turns away from the new 'Barbaric England' he has presented to 'contemplate the past before all is lost to the vandals' (Hoskins, 1955, pp. 233–5). Hoskins chooses to tell of his personal 'small' view, describing the landscape seen from his room in Steeple Barton in Oxfordshire, on the lower dip-slope of the Cotswolds. Reading, being an activity pursued alone, is again an appropriate metaphor for this telling of a personal landscape. In Hoskins' telling the personal, an even localler local, emerges too as a counter to the modern and its standards and plans, a 'life-world' opposed to a 'system'.

Hoskins begins, 'The view from this room where I write these last pages is small, but it will serve as an epitome of the gentle unravished English landscape.' Closed off, 'circumscribed' by the tall trees a half mile away, Hoskins' view 'contains in its detail something of every age from the Saxon to the nineteenth century'. Hoskins guides the reader gently over this portion of the parish, over his house itself, rebuilt 'over and over again' since the first residence on the site was erected in 1215, over the trees, the sites of mills, the stream, which one sees, the water cloudy after rain, 'as the Saxons saw it a thousand years ago', over the hedges signalling the Enclosure Acts, and over a small park with 'modest landscaping' dating from the 1870s (though the estate itself has documented history to Saxon times) . . . 'here in this room one is reaching back, in a view embracing a few hundred acres at the most, through ten centuries of English life, and discerning shadowy depths beyond that again' (Hoskins, 1955, pp. 233–4). Then, 'By opening the window and leaning out, the parish church comes into view', rebuilt in 1300, and now isolated. Out of Hoskins' sight, Hoskins says, next to his garden, is buried the main street of the village wiped out by the Black Death 600 years ago:

> One walks between the banks that show where the houses stood, marking how blocks of squared masonry thrust in one place out of the turf (a more important building than most of them) . . . one picks up pieces of twelfth-and-thirteenth-century pottery – mere sherds, bits . . . , but all detectable: nothing later than the Black Death, when the Great Silence descended.
>
> (Hoskins, 1955, p. 234)

One need only walk and look local to read all this and more, feet treading and mind and eye reading the centuries of England, picking up the details, piecing the place: 'Not every small view in England is so full of detail as this. . . . The cultural humus of sixty generations or more lies upon it.' Hoskins dates his view by personal life-span rather than the arbitary units of century and decade: 'But most of England is a thousand years old, and in a walk of a few miles one would touch nearly every century in that long stretch of time' (Hoskins, 1955, p. 235). And Hoskins ends his book with a motto to follow when abroad in the home country: 'Know most of the rooms of thy native country before thou goest over the threshold thereof. Especially seeing England presents thee with so many observables' (Hoskins, 1955, p. 235).

6. BLYTHE'S *AKENFIELD*

We now consider a book about a Suffolk village which is hard to categorise. Ronald Blythe's *Akenfield*, published in 1969, was already sold as 'a classic of its kind' on the jacket of the 1972 Penguin edition; but what 'kind'? Blythe's is a difficult book to place, a complex mix of social survey, oral history, personal vision. Penguin classified it as bridging 'sociology and anthropology'. Here we will elucidate its themes and its setting in the English culture of rural settlement. How does Blythe do his East Anglian English village?

Akenfield (the changed name of a particular Suffolk village north of Ipswich which Blythe studied) is a book made of typical figures. Save for Blythe's short Introduction, *Akenfield* is arranged by pieces on individuals, short descriptions of their looks, histories and occupations followed by their transcribed oral testimony to their life and places. The approach is similar to that found in much oral history, for example in Mary Chamberlain's own 1975 East Anglian rural 'portrait', *Fenwomen* (Chamberlain, 1983). The arrangement by type, the portrayal of a community laid out by its occupations, with individuals either not named or having their names altered for confidence sake, enables Akenfield to stand as a synecdochical place and book, individual yet having a wider implication. Its specificity gives off wider messages. And while earlier village surveys had contrived such synecdoche by reading places through predetermined categories of classification (Matless, 1991b), Blythe, by presenting his community through acutely personal detail, makes synecdoche appear to emerge from the village itself. In employing this strategy Blythe distinguishes his work from the procedures of social scientific survey, yet his particular synecdoche is in part enabled *by* the rhetoric of such survey. While the occasional chapter title in *Akenfield* has an air of self-conscious 'poetic' contrivance – 'The Ringing Men', 'The Northern Invaders', 'In The Hour of Death' – the more common division of chapters by straightforward name and occupational title – 'Ernie Bowes, thatcher', 'Marjorie Jope, district nurse', etc. – sees the author's hand retreat under the language of social survey, a terminology of occupations not specific to *Akenfield*.

Blythe subtitles *Akenfield* 'Portrait of an English Village'. While Hoskins employs a metaphor of reading, Blythe takes his cue from painting (the book is dedicated to the landscape painter and illustrator John Nash). The book is laid out in two dimensions, on one plane of the real. It is not an excavation beginning with appearances and digging for a deeper meaning behind. The testimony of individuals, and the observations of Blythe, makes up the place, gives it its reality. Blythe terms his book a portrait, but in many ways genre painting might provide a more appropriate metaphor for its way of telling. As in a nineteenth century genre painting, the picture of Akenfield is structured by signals which shuttle to and fro between and across chapters. The various stories speak to one another not through their sequence but more in the manner of glances across a room, across Akenfield, the book and the place. In each account Blythe gives the reader signs to other parts of the book and village, rather than attempting to bind the book in a linear account. Blythe of course presumes the reader to have an ability, by virtue of broad cultural familiarity, to read the signs, to make the connections. The reader is thereby flattered, or distanced, to some degree; either bound to, or set apart in culture from, the English village of Akenfield.

Akenfield is significant in its emphasising differences in attitude to the village and its environment between 'old' and 'new' villagers (Blythe's work comes of course a few years after the sociological enquiries on similar lines of Pahl and others (see Harper, 1989), though Blythe does not cite Pahl's work in *Akenfield*), and in its detailed treatment of what feelings and features define those contrasting groups. Different types of people are presented as reading Akenfield in quite different ways. Hugh Hambling, aged thirty, schoolmaster, whose account begins 'I am really a foreigner. I come from Norfolk', describes why he and his wife, 'village people two generations town-removed', moved, so to speak, 'back' to the country (Blythe, 1969, pp. 160–4). Hambling says he 'really . . . came here . . . to paint. . . . I wandered about seeing the land joining the sky in the huge way it does in Suffolk. I thought of Constable.' While on national service he still thought of this land and sky. And at art school 'nothing I learned hit me as this landscape does' (Blythe, 1969, p. 164). While this schoolmaster still finds Constable evident in the place, though, another informant tells Blythe of a Constable country lost. 'The Vet', Dr Tim Smith, fifty-five, remembers the place twenty years ago when he first came:

> All around, the country was as it always was – nice old parks, oak trees, rabbits, a few stags to make things look grand, sheep by the lake's edge, cattle up to their udders in the pond. That was how it was when I came here – England as it used to be. But it will never be a Constable again.
>
> (Blythe, 1969, p. 260)

For others still, though, those who constitute the 'native' population of the village, Constable doesn't even make an appearance in the landscape, doesn't crop up in their testimony to the place.

Blythe, presenting these different views in the various accounts, signals their presence in his Introduction, where he tells of two views of the village. The 'new villagers', who commute to the county and market towns, and who begin to 'claim' a village on arrival, seeking its identity that they might 'shape' themselves to fit it, and envying the 'old indigenous stock' yet having a life 'in effect . . . far freer than theirs', hold a view 'deeply coloured by the national village cult'. Akenfield holds for them ample 'evidence of the good life'; church, pub with local beer, stream, vicarage with cedar of Lebanon, school with tadpole jars, shops with doorbells, Tudor mansion, cottages and farms:

> Akenfield, on the face of it, is the kind of place in which an Englishman has always felt it his right and duty to live. It is patently the real country, untouched and genuine. . . . So powerful is this traditional view that many people are able to live in the centre of it for years and see nothing more.
>
> (Blythe, 1969, p. 17)

Blythe does not decry this vision, does not present it as somehow 'false', but rather regards it as a pleasure in place contrasting with a different pleasure in place felt by others. Of objections that village happiness is 'often exaggerated beyond all reason' he writes:

> Perhaps it would be fairer to say that two contrasting conceptions of this happiness, the new – i.e. the literate and informed – and the old – i.e. the mysterious and intuitive – are now existing side by side in Akenfield, and with scarcely any awareness of each other.
>
> (Blythe, 1969, pp. 16–17)

The old mysterious intuition does though come over in *Akenfield* as, if not a 'real' vision contrasting with the 'false' view of the 'incomer', then certainly as one possessing a gravity, a weight of reality, greater than that of the 'literate' tradition, a gravity grounded in a bond to the earth. Introducing a chapter telling of farm labourers, 'To be a Farmer's Boy?', Blythe finds 'the farm loyalist' unable to explain the attachment to agricultural work: 'One senses some more ancient pull for which there is no adequate sociological heading.' These '"opportunity"-resistant farm-minders' form that part of the community he terms 'the village proper. They are rooted-in deep and before they are middle-aged their lives have become entirely circumscribed by the parish boundaries' (Blythe, 1969, p. 80). The politics of such individuals, at the core of the village, also to Blythe defies the language of sociology. Recalling W. J. Turner on the politics of *Exmoor Village*, Blythe finds Leonard Thompson, aged seventy-one and a politically minded farm worker, 'untypically East Anglian, for politics on the corn plain are notoriously vague, furtive and unreal' (Blythe, 1969, p. 31). The issues pressing in Akenfield, and in like villages, are not the pressing issues of elsewhere. The 'villager', argues Blythe, whether farmer or farm-worker, is on one side of 'the great division which separates the growers from the mere consumers of food the world over'. Deep in their nature and 'elemental to their entire being' is an 'internationalism of the planted earth' – (one finds similar themes in the recent work of John Berger, and of course in many other writers before Blythe). This is no mere English rural spirit, but one shared with 'the rice-harvesters of Vietnam or the wine-makers of Burgundy'. All over the world, as here in Akenfield, is a surety and a stoicism beyond politics, a commitment: 'to certain basic ideas and actions which progress and politics can elaborate and confuse, but can never alter. Where the strict village existence is concerned it is Plus ça change, plus c'est la même chose' (Blythe, 1969, p. 15).

Blythe quotes a frustrated Welsh Rural Dean, the Rev. Gethyn Owen, in East Anglia since the war, who has long tried to fathom 'the great imponderable of the East Anglian character. . . . Fatalism is the real controlling force, this and the nature gods, the spirits of the trees and water and sky and plants. These beliefs seem to have no language, but they rule' (Blythe, 1969, p. 69). Akenfield's 'village existence' finds exaggerated form for Blythe in 'Davie' (no surname given), whose tale ends the book's Introduction. Davie's life has been 'a coda to the old existence' (perhaps Blythe is less confident of the unalterability of the 'basic ideas and actions' than he earlier suggested?). Taciturn, Davie 'likes to think that there is nothing to tell', and Blythe is puzzled by what keeps him going:

> Absence of commitment is a tremendous relief to him. . . . Yet the first thing one notices about him is not negation and recluse-like rejection, but an almost greedy, urgent positivity. It is such a vital existence which informs him. He is, he breathes – though what air? one sometimes wonders.
>
> (Blythe, 1969, p. 21)

He has lived in the same house for eight-one years, overlooking barley and a valley: 'This view is perfect and he hates it' (Blythe, 1969, p. 20). Davie is no Constable man.

In Akenfield's 'village existence' Blythe detects not only a local intuition but a local physique. People and their places are bound not only in emotion and intuition but, in this case at a regional level, in physical type. Dr Tim Swift, vet, may have thought that he wouldn't see Constable's Suffolk again, but he had only to look in the mirror to see an earlier local art reflected. Swift was, wrote Blythe, 'Tall, intelligent, with the red-brown skin and clear eyes of the "Ipswich" period Gainsboroughs (Blythe, 1969, p. 260). From Blythe's various comments emerges a photo-fit of a typical East Anglian; bony faced (Blythe, 1969, p. 14), 'sharp-featured and fair' (Blythe, 1969, p. 194), and 'tall, Viking-looking'. The latter phrase applies to Alan Mitton, aged thirty-eight, orchard foreman, whose name 'appears regularly in the church records way back to the late seventeenth century' (Blythe, 1969, p. 187) and who 'prays in lucid, mannered eighteenth-century English' (Blythe, 1969, p. 188).

Any slight inconsistency in the features mentioned does not matter here; with every mention of physique Blythe suggests a manifestation of the local type, that type being embodied especially in the blacksmith, Gregory Gladwell, aged forty-four, whose business is booming, due in part to the demands of newcomers:

> He is lanky and fine-boned – not at all like the classic forge Samson. Silky black hair fringes out from an old hat and the shirt gapes to reveal an ascetic olive-skinned chest stippled with spark-burns. In some way Gregory could be a Celt, yet not with those tall gaunt lines of nose and cheek; the eyes piercing their way from their deep sockets. These are East Anglian features, to be traced in the Ipswich or Norwich streets or in stone high up on the church walls.
>
> (Blythe, 1969, pp. 109–10)

Gregory Gladwell, at his anvil, displays local man.

Beside such local continuities of physique and intuition, Blythe also presents the persistence of a primitive agrarian culture, with activities taking place around the events of the seasons, activities which suggest Akenfield as possessing at its rural heart not so much quaint custom than an earthy culture of quite basic, at times 'base' behaviour. A key instance here occurs in the tale of Michael Poole, an orchard worker. In the chapter on 'The Orchard Men'

Blythe records the harvest activity of twenty years ago, since which time, as he makes clear in the story of Alan Mitton, the orchard foreman, procedures for gathering the crop haven't changed a great deal. Then, as now, groups of women would join the 'Orchard men' to pick. And on such an occasion, Blythe recounts, Michael Poole, now thirty-seven, lost his virginity. Illiterate, 'simple' in the terminology of others, 'sharp-featured and fair' and with a face showing 'the alertness of a forest creature' (Blythe, 1969, p. 194), Poole tells Blythe of the occasion when at the age of sixteen an older and larger woman who was picking the harvest came across the orchard one morning and 'She got her mouth on my face and, my God, she must have thought it was her breakfast, or something' and then later she came back and came 'down on me like a ton of bricks. I couldn't see nothing but grass. There was such a rocking. I couldn't tell whether I was babe or man.' The woman shrieked at him with her friends when she left at tea-time: 'It was my first time. Christ, that was a summer and no mistake' (Blythe, 1969, p. 196). Such things, Blythe implies, do happen in Akenfield. In recounting such incident, in making agrarian Akenfield the site of Poole's lost innocence, Blythe breaks any vestige of an innocent pastoral which scenes of apple-gathering might have conjured in the mind of the reader imagining a Suffolk summer. Some certainly might find the telling of such events just north of Ipswich at odds with their vision of the English country. (Sexual themes are of course common in the literature of pastoral and rurality, including in twentieth century English rural writing. Think of the novels of T. F. Powys, H. E. Bates, D. H. Lawrence. There is wide variation in this, though, from the erotic to the boisterous. On the English rural novel see Keith (1975), Cavaliero (1977).)

A similar effect of possible idyll-disturbance proceeds from Blythe's wider accounts of local agricultural process. Just as agriculture is unmistakably portrayed at the heart of the village, so Akenfield's is a heart not untainted by brutality, by the motives of profit and by the deployment of the chemical. The testimony of Dr Tim Swift, vet, contains a blunt account of the brutalities of factory farming in the village, of castration without anaesthetic, of thousands of chickens 'all deranged' in batteries, of cannibalism by chickens and pigs: 'The imprisoned creatures are eating each other. Everything is being controlled except their natural instincts' (Blythe, 1969, pp. 262–4). By operating against 'nature', human action is shown to merely give vent to the harsher instincts of that nature. Blythe too does not ignore the extensive use of chemical fertilisers in Akenfield's agriculture, though here, in his Introduction, agriculture, and the 'countryman' engaged in it, remain the 'real' and the 'elemental' in spite of 'science' and 'drugs'. Blythe has a faith in the continuity of earth and the human despite progress, comparing the earth's response to science in farming to the 'countryman''s response to modern life:

> He has his 1960s comforts and luxuries, as well as a fair inkling of popular sixties culture, but these things, though grabbed for by one who has a long memory of bleakness, are apt to be regarded as trimmings. The earth itself has its latest drugs and fertilizers poured into it to make it rich

> and yielding, but it is still the 'old clay'. In both its and his reality, the elemental quality remains uppermost. Science is a footnote to what he really believes. And what he knows is often incommunicable.
> (Blythe, 1969, pp. 14–15)

'Progress' hasn't (yet) eroded the country. Whether the 'villager' be an 'old horseman' or a 'rich agricultural technician for whom the word "farmer" is beginning to sound a quaint description', 'both will be one in the great division which separates the growers from the mere consumers of food the world over' (Blythe, 1969, p. 15). Agriculture and the farmer then remain for Blythe, with their brutal and chemical action, the core of the rural. In the following years though, as will be discussed below, it would be precisely such actions which would lead agriculture to lose its standing as the culture of the earth, to forsake, in the hearts of many, its place at the heart of the country.

In keeping with his emphasis on the use of land, Blythe presents his Akenfield not only as a village but as a parish, as a settlement bound to its land. In his first sentences Blythe portrays its situation, off an old straight Roman road ('the kind of road which hurries one past a situation') and folded away into its land, out of most sight, a place which 'Centuries of traffic must have passed within yards of without seeing', a place 'folded away in one of the shallow valleys which dip into the East Anglian coastal plain' (Blythe, 1969, p. 13). The emphasis on the parish, a unit both administrative and ecclesiastical, serves to bring together not only Akenfield and its land but also Blythe's secular and spiritual themes. Blythe consistently describes Akenfield with a reverence; this is a straightforward account which regularly hints at purposes beyond naturalistic record, purposes almost of worship. Of the parish church Blythe writes of how it 'retains the mysterious quality of an ancient sacred place which has never been out of the possession of a long line of simple rural people' (Blythe, 1969, p. 59). The church and the chapel though, Blythe suggests, are not the only sites of worship in Akenfield. A teacher at the local Agricultural Training Centre tells of how the older people 'communed with nature': 'They are great observers . . . they recognize a beauty and it is this which they really worship. Not with words – with their eyes' (Blythe, 1969, pp. 63–4). Blythe himself suggests a wider worship, a reverence for the parish as a whole, by placing lines from T. S. Eliot's 'Little Gidding' at the head of his chapter 'God':

> If you came this way, . . .
> It would always be the same . . .
> You are not here to verify,
> Instruct yourself, or inform curiosity
> Or carry report. You are here to kneel
> Where prayer has been valid.
> (T. S. Eliot, quoted by Blythe, 1969, p. 57)

Blythe too acknowledges an element of worship in what he presents as the 'national village cult' informing the preferences of incomers. After detailing the many elements of the 'cult' (see above), Blythe's tone shifts from near-satire to respect: 'It is patently the real country, untouched and genuine. A holy place, when you have spent half your life abroad in the services. Its very sounds are formal, hieratic; larks, clocks, bees, tractor hummings. Rarely the sound of the human voice' (Blythe, 1969, p. 17).

The 'cult' begins to take on the gravity of faith, the village becoming a sacred and revered whole. And whether from this or any other viewpoint, it is very much as an enduring whole that Akenfield is presented by Blythe. In his Introduction he says how 'Jets from the American base at Bentwaters occasionally ordain an immense sound and the place seems riven, splintered – yet it resumes its wholeness the second the plane vanishes. Nobody looks up.' Blythe asks, 'Could this be village indifference or village strength?' (Blythe, 1969, p. 14). Whichever, Akenfield sits stoical and one, an English village under the passing modern, and in this brief morality tale American, sky.

Blythe gives his own survey of Akenfield a hint of the sacred when he presents his book as 'the quest for the voice of Akenfield, Suffolk, as it sounded during the summer and autumn of 1967' (Blythe, 1969, p. 18). Blythe indeed locates part of the village voice in the Akenfield wind itself: 'the East Anglian wind does far more than move the barley: it is doctrinal. Probably no other agent . . . has done more to shape the character of the people' (Blythe, 1969, p. 21). The reverence for place embodied in Blythe's quest finds its most sustained expression in the chapter 'Not by Bread Alone', the testimony of 'The Poet'. Could the unnamed 'Poet' be Blythe himself? The Poet tells of the aptness to his or her vocation of cultivating a patch of land: 'So much of poetry is oblation and the putting of the seed into the ground is also a religious rite'. When living in London the Poet had written 'a poetry of despair', a 'continuous cry' of loss for the country: 'I came to this village to find my health. My wholeness. . . . Words have meaning for me here. . . . I came here to get better but I have in fact been re-born. I have escaped into reality.' This 'reality', more solid and of deeper meaning than the city, is a place of names, a seen, whole place: 'There are no nameless faces; I am identified and I identify. All is seen.' At Akenfield, 'the deep country', the Poet finds too 'the power of wonder. . . . In spite of machines and sprays, I still find Nature with a capital N in this valley.' And to be in Akenfield is to realise oneself as a part of this Nature:

> City life fragments a man. He is not complete when the reminders of the great natural complex of which he is a part are absent. The business of poetry is to mend the fragmentation which occurs when men forget their place in the natural creation. City poets are in danger of blocking the imaginative river with concrete and hearing so much noise that they miss the voice of the Goddess!

Blythe allows to speak a rural poetic aesthetic of wholeness, identity and oneness with the world, natural and human. 'The city poet' instead is one who 'records an alienation'. For the Poet the 'seasonal design of country time' is a counter to the twentieth century's 'immense alienation experiences'. Time is different in a village, slow and natural, and 'its poetic value' has been revealed to the Poet by Akenfield (Blythe, 1969, pp. 266–8).

One cannot read 'The Poet''s thoughts as necessarily matching Blythe's own, but one can I think project much of the feeling expressed in this record of a testimony on to its recorder, in particular the phrasing of village versus alienation. Akenfield comes over as a place intimate with its land, and perhaps above all as a place of names. The whole book is a gallery of (changed) names, full of characters. Akenfield becomes personalised and familiar, for its residents and for its reader, a personalised synecdoche of a place, an English village typical yet individual. Akenfield's wider message is perhaps to moot to its readership, whether rural or urban, that all English villages might be at once so personal, so singular, and so typical.

7. ON COMMON GROUND

i. The changing scene

Under scrutiny here is *The Living Landscape* (1986) by Fraser Harrison, a collection of essays on Harrison's then home village of Stowlangtoft in Suffolk, and the work of an organisation established in 1983 and of which Harrison is a prominent supporter, Common Ground. Again we find a complex bag, a landscape of morality, ecology, pleasure, the aesthetic and the spiritual. Harrison and Common Ground's work is given considerable space here, not least because of this complexity. In it we find expressed many of the changes in the imagined landscape in the past twenty years, and, particularly pertinent regarding the purposes of this essay, sustained reflection on issues of representation in relation to both landscape and place.

Harrison's and Common Ground's work both look to and in significant ways differ from the earlier writings of Blythe and Hoskins. The relation to Hoskins is considered below. Between Akenfield and Harrison's landscape though lie seventeen changing years. Both Harrison and Sue Clifford and Angela King in their 'Action Guide To Local Conservation' produced for Common Ground, *Holding Your Ground*, cite *Akenfield* as part of their lineage. Clifford and King cite Blythe's book as one of a number of 'Parish Books' to which their readers might make reference (Clifford and King, 1987, p. 205), while Harrison cites it in similar vein as an example, the most recent, of a 'descriptive tradition' in English rural writing (Harrison, 1986, p. 7).

Geographical description, the means to it and its end, lies at the heart of Harrison and Common Ground's work. Description is presented as an act possessing both use value and virtue in itself, in the attention to place and detail engendered, and in the subject-matter described; landscape, place, nature, home. For Harrison description becomes a moral, even spiritual task; naturalism is raised above itself by its object. Harrison presents descriptive rural writing as a tradition lapsed, Blythe's *Akenfield* being the last example. Introducing *The Living Landscape* Harrison plots 'the tradition of observing and celebrating the countryside for the benefit of a largely urban readership', a tradition beginning for him with Richard Jefferies in the late nineteenth century and peaking in quantity, though not always in quality, in the 1930s and 1940s (Harrison, 1986, p. 1). Harrison notes with a certain frustration the great boom in reissuing such work in the late 1970s and 1980s, a trend for him without discrimination regarding literary quality, such that any piece by Jefferies, W. H. Hudson or Edward Thomas, 'the giants of this new and ghostly publishing category', becomes destined for anthology, regardless. Edith Holden, with her best-selling *Country Diary of an Edwardian Lady* and its spin-offs, sits as 'the queen of such disinterred authors' (Harrison, 1986, pp. 2–3). Harrison detects a paradox. On the one hand the success of a work such as Holden's he regards as indicating 'a great force of feeling in the public, which was released, rather than stimulated, by the Diary' (Harrison, 1986, p. 4). Harrison records an intensification of emotion for country and wildlife since the mid-1970s, expressed in increased visits to the country and to National Trust properties, in the upsurge in membership of the Royal Society for the Protection of Birds, and in 'the insipid romanticism of television commercials exploiting the notion of "country goodness" . . . the near-terrorist activities of certain protection groups . . . the compulsion to turn our homes into farmhouses by filling every room with stripped pine furniture . . . the wildest extremes of eco-freakishness' (Harrison, 1986, p. 5).

On the other hand though Harrison sees at the same time a dearth of creativity, an absence of contemporary 'country writing' or 'landscape and nature painting' of 'outstanding worth'. While all around the country has emerged as a great field and hedgerow of care, the 'most important' branch of 'rural literature', the 'descriptive and observational' tradition, has for Harrison declined (Harrison, 1986, p. 6). Harrison cites with approval the influence of such academic works as Raymond Williams' *The Country and the City* (1975) and Howard Newby's *The Deferential Worker* (1977), the conservation writing of Richard Mabey, and the stories and essays of John Berger in *Pig Earth* (1979) recording life in a French peasant village (Harrison, 1986, pp. 6–7), yet he can find no works of comparable stature in the English descriptive vein. This lack is for Harrison no mere literary matter, though. This 'gap in our country literature' means that despite increasing knowledge of rural history, sociology, topography and ecology, 'we no longer know how to respond emotionally and philosophically to nature' (Harrison, 1986, p. 8). The descriptive and observational tradition, Harrison would argue, gave in both its content and in the labour of its writing a contact with nature and landscape. This was a literature dependent upon the author's proximity to place, and Harrison seeks in his own writing to restore this ethic of engagement and regard. In *Second Nature*, a volume edited by Richard Mabey for Common Ground, Harrison, writing on 'England, Home and Beauty', portrays his own as an earthy art, a literature of the soiled keyboard: 'As I type these lines, my hands are still dirty from digging up honeysuckle (Ionicera flexuosa, its label informs me)' (Harrison, 1984, p. 166).

Harrison cites *Akenfield* as 'perhaps the last book to be written in the descriptive tradition', and applauds its 'implicit discussion' of the relationship between people, land and landscape. Yet *Akenfield*, to Harrison, sits in another age:

> Although the book was only published in 1969, it has an old-fashioned air in that it is not overshadowed by an awareness of the ecological crisis, and it belongs, therefore, to another era of 'country writing', quite distinct from the present one.
>
> (Harrison, 1986, p. 7)

Since *Akenfield*, Harrison asserts, an 'age of grief' has dawned upon us; has come to pass, and come to our notice. Shortly after 1969, 'the public first discovered the massive scale of ecological decline' (Harrison, 1986, p. 7), a decline such that reading the reissued works of earlier writers, even those portraying the grim life of the rural poor, 'cannot help but remind us of a vanished landscape which was more beautiful than our own' (Harrison, 1986, p. 3).

Harrison cites the ecological decline as an aesthetic decline, and indeed his and Common Ground's work is notable for its assertion of ecology as a moral and aesthetic as well as a scientific concern. Harrison describes the 'decline' in the language of religion rather than of science, describes a fallen Suffolk in an 'age of grief'. Such a diagnosis demands a response less technical than imaginative, less of expertise and legislation than of a faith in nature and the aesthetic capacity of a human response. Harrison values the honeysuckle's dirt above its scientific label. Clifford and King, introducing *Holding Your Ground*, criticise how 'the "experts" have monopolised the discussions and decisions about our environment. . . . Science has tended to devalue the spontaneous response of the senses to nature, landscape and place' (Clifford and King, 1987, Introduction). Richard Mabey, in *The Common Ground*, the book from which the body took its name, likewise asserts 'unquantifiable values', arguing that questions of conservation 'cannot be answered by ecologists and agriculturalists alone, nor by attempting to reduce what are essentially moral problems to scientific ones' (Mabey, 1980, pp. 50, 25).

In *The Living Landscape* Harrison argues that while 'the biological case for conservation' is 'indisputable', 'the cultural case has not been put at all'. No one, he suggests, has 'demonstrated exactly' the spiritual 'human loss' following 'losses in nature', and this he regards as a moral as much as an intellectual failing: 'All of us, not just conservationists, are suffering from a lack of values: we face a moral, no less than an ecological crisis' (Harrison, 1986, p. 15). The vocabulary of recent loss is of course nothing new in this genre of writing, indeed it is as old as that writing. In *The Country and the City*, the work cited by Harrison, Raymond Williams writes of an historical 'escalator', which would seem never to have stopped, whereby successive generations of writers have located beauty and a settled agricultural way of life as only recently passed (Williams, 1975, pp. 18–22). With regard to issues of conservation the language of threat and loss became especially dominant from the late 1970s.

Many works portrayed a human 'heritage' of architecture and landscape 'in danger' (Wright, 1985; an important example is Cormack, 1978). Scenic and architectural beauty, and a particular way of agriculture, which will be returned to below, are similarly on Harrison's escalator, but in his work another passenger appears, one seldom lamented in isolation before; Nature. This is not merely a lament for the passing of an odd species or the destruction of a rare specimen; for Harrison Nature as both an ecological whole and an imaginative concept is threatened in this 'age of grief'. And as one form of agriculture passes from view on the escalator of history, so another, 'agribusiness', comes to the fore as the villain of the piece. A possible enemy of landscape for Hoskins in 1955, agriculture is now an enemy of Nature. Richard Mabey, writing in 1980, records how:

> The greatest shock in the present transformation is that it has come about not so much from an invasion by urban sprawl or industrial development, but from insidious and often unobserved changes in the internal workings of the countryside itself. We were not prepared for this. The attacks upon the countryside that accelerated in the expansionist years after the Second World War had made the conflict seem a very clear and traditional one. It was the green fields of St George's England versus the dragons of mammon and industry.
>
> (Mabey, 1980, p. 22)

Now, however, there appears an enemy within. In 1986, Harrison begins *The Living Landscape* in taking this change as read: 'As everybody knows, our national heritage of nature is steadily being diminished.' This 'tragedy' is for Harrison such that 'any description of the countryside is bound to be an exercise in either anger or sadness. No other tone is appropriate.' In this age, 'no writer today can blithely pipe songs of pastoral joy while the plough hisses through virgin meadows and the power-saw snarls over its woodland prey'. Modern agriculture is in this to Harrison committing a cultural as well as ecological offence:

> by silencing the old music which spoke of a carefree enjoyment of rural nature, modern agribusiness has done far more damage than merely cutting down the expressive range open to country writers; it has also threatened a crucial part of our national sensibility. As much as the innumerable losses to our wildlife, this loss is the tragedy of our time.
>
> (Harrison , 1986, pp. 1–2)

Agriculture's passage from custodian in the country to villain of the land is indicated by the grouping in *Holding Your Ground* of 'Farming and Pollution' in the same chapter (Clifford and King, 1987, pp. 165–81). Farms, rather than making up the English landscape, sit as loci of potential hazard, whether observed, as in the case of plough or saw, or, and this would seem harder to grasp within an observational tradition, unobserved, as with the invisible workings of the chemical. It is indeed interesting that Harrison, with his language of aggressive destruction and violation, focuses on the very visible agents of change – the plough, the chainsaw. Though the results of pollution – a lack of fauna and flora for example – can be and often are described (negatively) in visual terms, and while for example a crop-spraying plane might be highlighted as a threat, the 'insidious' operation of the chemical defies observation. One could perhaps argue, again quoting Mabey, that regarding the traditions of landscape description, linguistically as well as imaginatively, 'We were not prepared for this.'

Paralleling this condemnation of the practices of contemporary farming, Harrison makes a key claim to an observer's rights over nature and landscape. As nature is a key element of national culture, so for Harrison the ownership of nature is as much vested in the observer as in the land-managing farmer. The observer has for Harrison a real claim to landscape, one which should be built into agricultural and conservation policy, and the category of 'observer' includes not only the visitor but the greater part of those living in the country. Harrison disputes an equation of interest between 'farming' and 'countryside', and claims a local right which comes from residence (and not necessarily for a long period) rather than from any economic engagement with the land. He is, in the title of Mabey's introduction to *Second Nature*, 'Entitled to a view', in both senses of that word:

> I do not work on the land, or own any of it, or play any part at all in the processes which shape it. Yet I am not a tourist, and I belong to this patch of countryside as surely as the earthiest farm labourer. I live and work here, and have done so for ten years, which is, incidentally, longer than many other inhabitants.
>
> (Harrison, 1986, p. 12)

Harrison and his family utilise local education, shops, pubs and garages:

> I am no different from the great majority of my neighbours, who are not farmers either. . . . But, like the rest of the rural population, we are alienated from our own countryside, . . . none of us has any control or influence over what happens to our surroundings. . . . We can only look on helplessly . . . we are being turned into voyeurs.
>
> (Harrison, 1986, p. 12; for a similar expression of helpless detachment see Mabey, 1980, p. 20)

Harrison proposes a way out of what is for him an undesirable state in his essay 'Who Owns Nature?', where he attempts to pre-empt voyeurism by claiming a form of ownership over the landscape. Lamenting the increasing concentration of ownership in fewer hands, 'whose activities are not easily controlled by law, whose methods are inimical to conservation, and whose handsome income is derived from public money' (i.e. whose activities are defined as against a 'public' interest), Harrison claims a collective cultural ownership of the countryside, and of the nature which that countryside contains: 'although nature may, in the strictest sense, belong to nobody, the countryside, which is an entirely social and cultural concept, belongs to everybody. In short, it may be their land, but it is our countryside' (Harrison, 1986, p. 75).

A particular notion of observation is a key to enabling such an analysis. Observation is presented as an active engagement with place and landscape, an active moral and aesthetic endeavour distinct from the 'voyeurism' of a passive onlooker. And observation also stands as an activity ennobled by long precedent, in contrast to the more recent plight of voyeuristic alienation. Introducing *Second Nature* Richard Mabey claims just such an observational tradition, pointing out that though the 'observers' whose writings and visual art make up the collection are none of them 'country' people, this is no departure from the traditions of rural writing: 'I suspect that much of what is generally regarded as the "rural tradition" has this repeated blending of foreign and local, old and new, right through its history. It is what has kept it alive' (Mabey, 1984, p. xi).

It is worth emphasising here that the writing under discussion here contrasts to some earlier branches of the 'rural tradition' in presenting itself as having a radical intent in opposition to what it sees as the established view. Mabey terms *Second Nature* as in a broad sense a 'radical' collection, encompassing ideas from the Left and/or the Green movement, and drawing on Eastern as well as Western philosophical traditions (Mabey, 1984, p. x). The volume contains pieces from a number of Left scholars who have done much to both foreground and interrogate ideas of landscape and countryside in recent years; John Barrell, Colin Ward, John Berger, Raymond Williams. The last three in particular have in different ways sought to combine socialist, anarchist and environmental thought to renew what has been presented as a tradition of non-corporatist Left thinking reaching back to Morris, Kropotkin, Thoreau and others. This is an intentionally subversive commentary on environment, which sees landscape as something worthy of aesthetic and moral attention, and for that reason, as well as for more directly political and economic reasons, lamenting and seeking to alter its present state. Harrison, also a contributor to *Second Nature*, writes in *The Living Landscape* of the aesthetic as possessing in relation to landscape an inherently subversive dimension (this in a way comes full circle from earlier analyses of Berger, Williams and others which saw the aesthetics of landscape as generally legitimising positions of authority). Harrison's observation is a subversive practice, his 'descriptive tradition' a naturalism not necessarily confirming the order described. Drawing on the vocabulary of trespass and poaching, 'traditional' crimes against the landowner often portrayed as a just exercise of denied rights, Harrison upholds a freedom to see: 'our aesthetic perceptions are quite insensitive to the patterns made by property. The eye cannot be confined or repulsed by gates and walls; it is an incorrigible trespasser and, to that degree, our aesthetic response to the countryside is always potentially subversive.' The farmer, says Harrison, may own the land, but 'does not own the view. Landscape is common property. . . . There is no land so private it cannot be poached by the inquisitive eye' (Harrison, 1986, p. 78) . . . the inquisitive and observing rather than the passive and voyeuristic eye.

The criticism of modern agriculture in *The Living Landscape* and *Second Nature* is not confined to the aesthetic. In both books demands are made for a restructuring of the rural economy, and a redistribution of property. Harrison writes of 'the key to conservation' lying in 'restructuring the economics of the farming industry', reducing the protection given to large farmers in order to produce 'a radical change in the structure of ownership' (Harrison, 1986, p. 75). And in *Second Nature* Raymond Williams seeks to broaden arguments on rural economy and society, looking to a future without large-scale capitalist farming, with ecology and economics combined in a 'green socialist' ('the most hopeful social and political movement of our time') idea of 'livelihood' (Williams, 1984, p. 219). The detail of such proposals is not our concern here, rather that they are being articulated within the broad project of Common Ground. This is not a vision of environment and countryside as an escape from the concerns of the contemporary world. That said though there is a definite element of yearning for a system of small farming portrayed as largely passed, an agriculture which rides on the escalator with nature, the latter's presence there being seen as due in part to the former's demise. To reclaim such a pattern of land use would for Harrison reform and restore 'the balance of ownership': 'The smaller farmer would return to the countryside, land would be less exhaustively cultivated, and the consumer would benefit no less than the environment' (Harrison, 1986, pp. 75–6). This would be an agriculture following the precepts of ecology: balance, equilibrium and renewal, in itself and with the wider rural economy.

Again the detail of any proposals for achieving such an aim is not our concern here. Indeed in the work of Common Ground and of Harrison such proposals are not to the fore. The above discussion has rather served to sketch the change of discursive scene between *Akenfield* and *The Living Landscape*; the emergence of a concern for Nature in itself, the criticism of modern agriculture, the criticism in Common Ground's work of a purely 'scientific' conservation, and the claim to an observer's right over landscape which comes increasingly into play in discussions over planning and development in the countryside in this period. We now turn to examine in more detail the idea of settlement, in particular rural settlement, articulated in the work of Harrison and Common Ground, an idea revolving around particular notions of 'the rural', of locality, of home and the everyday.

ii. Settlement

a. The rural and the parish

Despite the fall of Nature, despite the alienation Harrison feels in the contemporary country, and despite the reasonable insistence of Common Ground that their ideas apply as much to urban as to country places, the rural still stands, in their and Harrison's work, as a site of potential and actual redemption; a site of landscapes and the organic, and of visible places and parishes. In *The Living Landscape* Harrison indeed calls specifically for, and presents his book as a humble contribution towards, a 'new rural aesthetic', a required response to the 'moral' as well as 'ecological' crisis facing contemporary society: 'We lack an aesthetic which, on the one hand accommodates the awful fact that nature is being killed off, and, on the other, expresses once more our age-old sense of spiritual or imaginative identification with it' (Harrison, 1986, pp. 8–9). Harrison offers his 'poetic testimony', his 'witness . . . to the feelings, and . . . the ideas aroused in me by my landscape' (Harrison, 1986, p. 15), to a new rural aesthetic which might 'extend our range of feeling' beyond current 'mourning or recrimination', and would transcend the 'age of grief': 'Without indulging in fatuous romanticism, we must achieve a new communion with nature and open an exchange which expresses the full span of our emotions' (Harrison, 1986, p. 15).

Such a projected spiritual response demands of course the identification of an existing lack, a moral hole to fill. Drawing on Keith Thomas' *Man and the Natural World*, Harrison traces such a lack to the sixteenth century, the time of a shattering of union: 'I am convinced that our divorce from nature, though an inestimable boon in setting us free from laborious toil, has nevertheless added to our sense of spiritual emptiness' (Harrison, 1986, p. 48). Harrison traces 'the story of our alienation, to use a twentieth century word', a break from nature completed by nineteenth century science. Now, 'like it or not', we are an 'urban industrial society', and the task becomes 'to hold the line of damage . . . it is consolation in the here and now that we require' (Harrison, 1986, pp. 50–1). In introducing *The Living Landscape* Harrison, reminiscent of Blythe's Akenfield 'Poet' (and Stowlangtoft is not far from 'Akenfield'), indeed details his own personal escape from the here and now of urban industrial London to the consoling Suffolk village he writes in (Harrison, 1986, pp. 9–11). Stowlangtoft, and the rural as a whole, is presented by Harrison as providing the material for consolation, the imaginative resources required to weave a new engagement with the natural, perhaps leading to a future renewed marriage.

Harrison regards the collective divorce from nature as running in tandem with a general decline in a 'matrix of tradition'; ethical absolutes, and accepted norms relating to family and religion (Harrison, 1984, p. 169). Under these circumstances, Harrison writes in his *Second Nature* piece on 'England, Home and Beauty': 'the need to preserve our wildlife and countryside is nothing short of imperative, for no other body of symbolic reference can fill as much of the void left by the collapse of this connective tradition as can rural nature' (Harrison, 1984, p. 169). The rural, with its nature and symbol, can fill an ethical, moral and spiritual hole, one which, if left empty, might endanger 'civilised life' and 'collective sanity':

> It is my belief that we, as a species, have an indispensable need of an intimate and harmonious relationship with nature, and that if we are deprived of that relationship we will be quite unable to achieve the level of personal and communal wellbeing which is necessary for civilised life . . . prolonged divorce from nature and the emptying-out of our age-old symbols will ultimately drive us into collective madness.
>
> (Harrison, 1986, pp. 14–15).

Harrison's themes of morality and spirituality are of course common in the rural 'descriptive tradition' in which the author places himself. Richard Jefferies, for Harrison that tradition's originator, is an obvious example. One of the more curious and interesting elements of Harrison's argument though is his assertion of myth and spirituality not as a driving rock of faith but as an 'illusion' providing 'consolation'. This language of consolation and illusion is also to the fore in Harrison's writing on memory and nostalgia. Nostalgia, 'the making of memory-myths', plays for him a 'vital part' in the making of individual and collective identities: 'Illusions, in a word, are essential to our imaginative wellbeing . . . illusion is indispensable to conceiving all kinds of models for improvement, whether religious, ethical, political or social' (Harrison, 1986, pp. 109–10).

In his essay on 'Positive Nostalgia' Harrison conceives of such illusions as being at the heart of pastoralism, a mode of thought offering both a relief and possible escape route from the 'continuing estrangement from nature' he terms 'a virtual guarantee of psychopathy on a national scale'. Nostalgia and the pastoral have at their heart a vision of 'a better way of life', of harmony, communality and affinity with nature. For Harrison the fact that the English pastoral is 'conjured out of material inaccurately derived from rural history and certain persistent myths about Old England, should not detract from its worth as a dream of an improved, kindlier and more connected existence' (Harrison, 1986, pp. 120–1). And this is a dream which can evidently find its most appropriate location in contemporary village England. In response to those writers (and Harrison may be thinking of Martin Wiener's 1981 *English Culture and the Decline of the Industrial Spirit* here, as well as anticipating Robert Hewison's 1987 *The Heritage Industry*) who regard pastoralism and nostalgia as symptomatic of and aggravating a collective malaise, Harrison states that: 'a society which is addicted to nostalgic longings for a lost and largely invented countryside is probably sick, but a society which feels no longing for its countryside is sicker still' (Harrison, 1986, p. 121).

Harrison is notable, though not alone, in extending ideas of illusion and consolation to cover that least whimsical of realms, the spiritual. Harrison declares himself an atheist (Harrison, 1986, p. 61). Writing of feeling 'soothed and heartened' whenever he enters Stowlangtoft church, Harrison says: 'No other building I know has the same power to restore. However none of this is to imply a dawning of faith.' Harrison's habitual pursuit of 'church crawling' has rather 'confirmed' his 'lifelong atheism', though in this pursuit he is less a 'mere tourist' than a 'secular pilgrim' (Harrison, 1986, p. 61). Beyond the church, in the landscape, Harrison similarly juggles secular and religious language to describe his response to nature. Just as the Suffolk church arouses in him emotions 'complex and powerful', which he ascribes to 'a direct response to the religious essence of the church and its symbolism. Or rather, they are a response to those great themes of existence – birth, marriage, morality and, above all, the mystery of death – which are the special province of all religions', so nature, though no longer regarded as 'the handiwork of God' but rather as 'its own creation', provokes for him a 'religious' response: 'we still retain a "religious" perception of landscape, that is a capacity for discovering in nature a symbolic representation of our deepest concerns and aspirations' (Harrison, 1986, p. 35). Harrison later writes of how: 'In the absence of traditional forms of idealism and moral discipline, which are provided by communal religions deriving from belief in a superhuman authority, we must devise a humanistic ethic' (Harrison, 1986, p. 96). Harrison's then is an atheism and a humanism based less on rationality and secularism than on a search for common 'human' emotions embodied in a 'human' response to nature, emotions which might comprise an equivalent for the spiritual.

Harrison also proclaims himself a socialist, and indeed is not the only writer of the Left to address themes of nature, the spiritual and the moral in recent years. John Berger is another (for example Berger, 1985, 1986); though more closely paralleling Harrison is his friend and accomplice in church crawling, the late art critic Peter Fuller, whom Harrison acknowledges in *The Living Landscape*. Fuller himself wrote of his friendship with Harrison in an essay on the English landscape artist John Piper, telling of their taking up the church crawling which Piper and John Betjeman had earlier practised (Fuller, 1985, pp. 92–3; Betjeman and Piper had also collaborated in editing the Shell County Guides). Fuller's essay sits in a book tellingly entitled *Images of God: The Consolations of Lost Illusions*, which also includes a piece on Neo-Romanticism subtitled 'In Defence of English Pastoralism'. Here Fuller, like Harrison, presents the pastoral not as escapism but as 'one of the few symbolic ideas in our culture from which we can draw some hope' (Fuller, 1985, p. 89). To Fuller, as to Harrison, 'A society which has no hegemonising religious beliefs lacks a shared symbolic order, with disastrous effects on both aesthetics and ethics' (Fuller, 1985, p. 88). Fuller too asserts his atheism; the language for his pastoral 'quest' is to be found not in past religious iconography but 'anew, in nature itself' (Fuller, 1985, p. 91; Fuller developed his arguments more fully in his final book *Theoria* (Fuller, 1988)).

In Fuller's work and in the journal *Modern Painters* which he founded, in Harrison's writing, in Patrick Wright's work, in the work of the contributors to *Second Nature*, in the activities of Common Ground and in the Green movement in general can be traced a widespread re-emergence in the 1980s of an English culture of nature and pastoral, and this often in the work of those critical of earlier manifestations of such a culture.

For such writers and artists it seems that nature and the pastoral is itself a site to be reclaimed and redeemed from its earlier cultural (ab)use.

To turn back to Common Ground and to Harrison's landscape. We have seen the role of nature in Harrison's redemptive rural. There is also the key and complementary role of place, and in particular place expressed as parish. Place, though, like nature, is dangerously on history's escalator. In *Holding Your Ground* Clifford and King write of the present as 'a time when regional and local distinctiveness are disappearing so fast'. . . . 'Fifty years ago it could be said that there were real regional and local differences in the landscape and in villages and towns' (Clifford and King, 1987, p. 2). The distinctiveness of place then is slipping, difference was there fifty years ago but is less so now (again one can trace many similar sentiments expressed fifty years ago); yet, as the whole project of Common Ground suggests, some identity of place is still thought to exist, to be maintained and celebrated. The celebration of place which Common Ground would subsequently promote was given a cultural prod by *Places*, an 'Anthology of Britain' edited by Ronald Blythe in 1981. Contributors included Blythe himself, Richard Mabey and Raymond Williams, along with many other poets, novelists, historians and photographers.

The volume was illustrated by the artist John Piper. Subsections of the book included 'Relics and Locations', 'In The Village', 'Borough Boundaries' and 'Personal Map Reference'. Places urban, rural, desolate, crowded, old and new were drawn on. The ethic and aesthetic informing *Places* would inform much of Common Ground's work, not least in the emphasis on the personal. Blythe writes of the essays in *Places* having the 'common theme' of 'personal geography', of 'an eloquent appreciation and defence of place in its most personal terms' (Blythe (ed.), 1981, vii). A promotion of personal connection and familiarity with a locality is at the heart of Common Ground's highlighting of place, in particular in their selection of the parish as an appropriate scale at which to discuss and celebrate. The parish sits as an imaginatively manageable space for the individual, as a scale at which it is possible to be familiar in detail with a surrounding. Mabey, in introducing Gilbert White's eighteenth century parish book, *The Natural History of Selbourne*, terms the parish 'the indefinable territory to which we feel we belong, which we have the measure of' (quoted in Mabey, 1980, p. 36; also in Clifford and King, 1987, p. 7). Clifford and King, writing of 'Personal Landscapes', moot that 'It is the size of the parish which makes us feel happy with it as our home range. . . . We love our home range because we know it, or parts of it, intimately' (Clifford and King, 1987, p. 4). At the level of the parish, it seems, one can cope, have 'the measure', and love all the more for it.

The choice of the 'parish' by Common Ground as the scale for discussion, rather than, say, the 'locality' or 'ward', again embraces a spiritual theme. The parish is the spatial unit in England which has both ecclesiastical and civil function, spiritual and secular meaning. And though the civil function of the parish may be relatively recent (generally late nineteenth century), the ecclesiastical purpose goes much further back. The parish can therefore be read as a unit where spiritual and secular histories interweave. To 'observe' the parish, as Harrison and others would urge, can suggest the spiritual and ritual as well as the ocular sense of that word. And at the centre of that parish is invariably presented the parish church. In a notable passage tying church, landscape and history, Harrison describes how in *The Living Landscape* he has

> tried to describe the way our village is rooted in both its history and its landscape by tracing the connections that bind the church and its iconography with its raw materials, oak and flint, and by showing how they all remain properties of something that is more an idea than a place – the parish. Through these materials and the cultural nexus they form, the continuities of time and place are made visible, immediate and, above all, tangible. The flow of human experience is made meaningful.
>
> (Harrison, 1986, p. 129; see also p. 66)

The parish becomes iconographic for those within, situating them with its tangible, visible meaning. And the church itself plays a key symbolic and topographic role in this: 'The church moves me too because it unifies the village and its landscape in a single, reciprocal creation' (Harrison, 1986, p. 61). Harrison cites its local materials, its Gothic symbolism drawn from nature, and its 'salient position on the skyline' which 'marries the community with its countryside, while pointing to transcendent aspirations' (Harrison, 1986, pp. 61–3). And this unity is for Harrison not one ordained by authority but one which expresses the popular: 'Every village church bears witness to another history, an alternative tradition of ordinary people that is not stamped with the badge of a single owner, or at least not as conspicuously as the landscape' (Harrison, 1986, pp. 64–6).

In a notable phrase Harrison writes of how the detail of the parish makes up an 'ecology of the imagination'. Landmarks, he says, 'cleave together in a kind of ecology of the imagination, which links farming with nature, property with beauty, community with landscape, and transcends them all'. Through this ecological 'imaginative coherence . . . the flow of experience is rendered meaningful' (Harrison, 1986, p. 78). The spatial unit Harrison chooses to make up his ecology is the parish, which in the imaginative sense becomes the unit of territory, the space of operation and allegiance. Tom Greeves, in a Common Ground booklet, writes of how Common Ground use the term parish 'to mean not only the conventional civil or ecclesiastical unit but also the idea of territory or neighbourhood of human scale, whether rural or urban, to which you feel allegiance' (Greeves, 1987a, p. 2). Again *The Common Ground* provides an antecedent, Mabey terming the parish boundary 'more the limits of our intimate allegiances than lines on a map' (Mabey, 1980, p. 36). Common Ground too has given extended attention to the parish boundary. *The Parish Boundary*, another booklet by Greeves, upholds the boundary as in turn 'tangible history', 'living nature', material for 'a celebration of the community', and 'an inspiration for rural skills, writing and art' (Greeves, 1987b). Greeves begins *The Parish Boundary* by quoting Hoskins from his 1959 *Local History in England*: 'you should walk around and describe the boundaries of the ancient ecclesiastical parish or the boundaries of whatever is your chosen territory . . . at least you will get the feeling of (it) in a way that nothing else can give'.

'How many of us', asks Greeves, 'have taken his advice?' Common Ground's work of the parish is here presented as reviving Hoskins' purpose of 'nearly thirty years ago' (Greeves, 1987b, p. 1).

b. The local

b(i). Detail and difference As much as Hoskins then, Common Ground upholds the local, in the form of the parish, in and for itself. What though is the detail of this parish aesthetic? Again, part of the detail is an emphasis on detail itself, for it is detail, rather than any composed overall picture, which is seen to make a place distinctive. In Common Ground's emphasis on 'Local Distinctiveness' (the title of their most recent campaign), Hoskins is again cited. Clifford and King, writing on 'Historic Landscapes and Ancient Monuments' and how to protect them, advise: 'Using W. G. Hoskins as your mentor, trace parish and manorial boundaries, old trackways . . .' (Clifford and King, 1987, p. 25). *The Making of the English Landscape* is termed a 'classic book': 'Anyone who cares for landscape and history cannot fail to be fascinated and excited at the insights and the questions this book raises' (Clifford and King, 1987, p. 24). Like Hoskins, Clifford and King employ the analogy of landscape as document to argue how: 'the fine grain of the landscape has been socially constructed . . . like old documents . . . [it] holds many keys to an understanding of our past, our present and the evolution of our culture' (Clifford and King, 1987, p. 6).

Common Ground's 1990 'Local Distinctiveness' campaign, aimed at 'Making the Difference', laments an 'erosion of difference and bleaching of identity, detail, craftsmanship and meaning' which 'affects us all, emotionally and culturally', and seeks in response to 'help people hold on to what is valued in the old, to demand the best of the new and to add to local identity'. Identity is then presented as an ongoing process, but before difference can be made, before the 'fine grain' can be accentuated, attention to that grain is demanded: 'first we need to recognise the differences and detail that enrich our places' (Common Ground, 1990a). It is worth noting that Common Ground does not restrict its attention to the visual sense in detecting difference. A recent campaign to save orchards gave the tongue a role, the maintenance of orchards being presented as intimately bound up with the perpetuation of local varieties. Roger Deakin, a founder-director of Common Ground, wrote in the *Financial Times* of how varieties of apple (6,000 registered in Britain but only nine dominating commercial orchards) brought 'a sense of local continuity' to the palate: 'tasting local fruit is oral history at its most enjoyable'. Deakin mourned the decline of a range of 'variations' on fruit themes that 'gardeners, nature and history' had invented (Deakin, 1989).

The argument for difference put forward by Common Ground indeed expresses a more general feeling that the modern world is one increasingly of standardisation and uniformity (again Hoskins sets a parallel). In *The Common Ground* Mabey portrayed the trend to 'uniformity' as a recent human trend against a nature characterised by intricacy (this is no violation of a 'simple' nature, indeed the human is portrayed as at present the more 'simple' element): 'What must worry us, now, is whether the overall drift of our society towards uniformity is already outstripping the contrary, natural drive towards intricacy. Change . . . can enhance intricacy – but only if its pace and scale do not exceed those at which nature can adjust' (Mabey, 1980, p. 47).

Mabey laments a trend against nature's ecology. Fraser Harrison though extends such an analysis to include the human cultural response to landscape: 'I fear the continuing expansion of a landscape which cannot reciprocate, which refuses the old interchange of feeling. . . . I fear the coming of a countryside which is so intensely farmed its landscape will be too barren to respond to our cultural needs.' Harrison, in a passage opposing his own view of beauty to a modern aesthetic of geometry (and, curiously given the history of both, opposing geometry and magic), writes of the physical landscape as losing its faculty to speak to the imagination:

> I have a nightmarish vision of a landscape drained of magic . . . I foresee a countryside stripped of poetry and supervised by a society which prefers the geometry of agriculture to the wayward tangle on its shrinking margins . . . a blank landscape confronting a blind nation. . . . This is how the light could fade: not eclipsed by the storm-clouds of nuclear winter, nor by the fog of pollution, but put out by the blinding of our imagination. Above all, it is the job and responsibility of conservation to keep alive the landscape's poetic faculty.
>
> (Harrison, 1986, p. 37)

A number of Common Ground publications direct their attention beyond the landscape to portray the wider trends of consumer society as being towards uniformity and standardisation. The parish becomes here a site resistant to such trends, from such resistance is seen to derive some of its potential. Tom Greeves writes of how: 'In an age of "Walkmans", multi-national companies and hypermarkets, the parish has survived as a unit that is both understandable and human in scale' (Greeves, 1987a, p. 3). Likewise the 'Local Distinctiveness' campaign leaflet announces the 'loss of distinctiveness' in terms of an advance of the modern standard, with the 'richness of local diversity . . . under siege. Mass production, increased mobility and forceful promotion of corporate identity has brought us uniform shop fronts, farm buildings, factories, forests and front doors. . . . New estates offer the "Cheviot" or "Purbeck" house in any part of the country' (Common Ground, 1990a).

This is a society with a cavalier outlook even on place-names. Local connection goes by the board of sale. Such pressures for uniformity are presented as often expressing a petty and unnecessary impulse to tidy. The bad side of the contemporary is one of interference in the local for no good regulation: 'A neat row of cypress trees replaces an ancient hedgebank; a pit tip is moulded out of recognition . . . ; an orchard becomes a housing estate, the stream culverted; . . . a wild old quarry is filled in with rubbish; footpaths are reduced to numbers on a map' (Common Ground, 1990a). Difference reduced for no sake. Common Ground's is not a plea for maintaining 'useless' idiosyncracy, but rather for resisting an erosion of difference for the sake of what are presented as almost the fussy whims of the contemporary to tidy away the local.

b(ii). The everyday, the ordinary, the unofficial, the personal Central to distinguishing between the good 'human scale' local and the bad forces of 'uniformity' is a particular notion of the everyday. The realm of the everyday and the familiar is upheld as one distinct from and at odds with (rather than informed and partly constituted by) the trends of the consumer world and the tendencies to the uniform; upheld as a life-world at odds with a system. On the opening page of *Holding Your Ground* Clifford and King reproduce what can count for Common Ground as an almost talismanic quote from Fraser Harrison's piece in *Second Nature*:

> The ordinary places and objects that make up our everyday landscape, our personal countryside, stand as living monuments to our continuing survival and feeling response to the world. Without such monuments, and they are not necessarily a rural monopoly, our sense of identity begins to crumble and warp. We need little, low, unspectacular corners which can carry special resonances for us alone . . . this complex intermingling of our emotions and their reflection in nature makes possible the birth of a powerful sense of rootedness and meaning in a world which otherwise yields little but confusion and futility. By conserving the mass of precious detail in our parishes, we conserve ourselves.
> (Harrison, 1984; quoted in Clifford and King, 1987, p. 1)

The recent 'Local Distinctiveness' campaign leaflet includes a proposal for compiling 'an alphabet of local distinctiveness', and gives an 'exploratory' example, from 'Ayrshires' and 'ammonites' through 'gasometer' and 'Norfolk beefing' to 'Yorkshire pudding' and 'Zennor' (Common Ground, 1990a). The alphabet is made up of things not only local and familiar but often self-consciously ordinary, things that wouldn't make it into most guidebooks, what Harrison terms in *The Living Landscape* 'the ordinary places and things that make up our everyday landscape' and that 'stand as monuments to our own existence in a way that the grand prospect never can' (Harrison, 1986, p. 130). These are less outstanding dramas standing out from a humdrum local than 'parochial monuments, landmarks . . . by which each person can take his or her bearings in time or place' (Harrison, 1986, p. 129). The 'Local Distinctiveness' leaflet, double folded, opens up once on to the exploratory alphabet reproduced over a drystone wall, and the second time on to written details and a photograph of the most plain street. This is though (I think) not meant as a picture of emptiness but as an image which says, 'you may not find anything special here, but were you to live here you would'.

In this espousal of the commonplace a textural rather than a visual metaphor is brought to bear. This is a feel for the place's 'grain', for the texture of place; less a spotting of highlights than a coming to know the intricate and intimate weaving together of things such that none are highlighted at a cost to others. The grain is to be maintained whole, not broken by a selection of the visually out-standing. There is again a parallel here with Richard Mabey's ecological arguments for conserving the common species. Mabey suggests in *The Common Ground* that 'Our historically conditioned view of nature', based on picking out the exceptional, 'has pushed conservation . . . into a potentially dangerous preoccupation with the exotic and the rare' (Mabey, 1980, p. 34). Mabey argues for attention to the common and the local rather than the exotic:

> In some ways, the local extinction may represent the greater overall loss, for here it is not just the species that is lost, but the day-to-day intimacy and associations, the neighbourliness, that builds up around a plant or

> animal that has lived on close terms with a human community. There is, I think, much to be said for using these human territories as the basic units by which one measures natural erosion.
>
> (Mabey, 1980, p. 36)

And Mabey indeed hits on a metaphor of texture to argue for an emphasis in conservation on incorporating change into continuity, rather than resisting change outright. Mabey writes of the

> peculiar combination of development and continuity, in which old forms are not so much replaced as contained inside the new, is characteristic of all living things, including ourselves.... It is like the development of grain in a tree.... I think that the metaphor of 'grain', with its suggestions that, where continuity is not broken, change can increase natural intricacy, can be applied to all natural communities and places.
>
> (Mabey, 1980, pp. 43–4)

Common Ground, then, has highlighted the familiar. The word 'highlighted' is indeed appropriate here, for the opposition of 'special' and 'everyday' is turned around by an insistence that the everyday is itself special. Clifford and King argue that because 'experts' have 'concentrated on the special things, the everyday surroundings of most people have been devalued by default.... Vernacular buildings, ordinary landscapes and their creatures ... embody the spirit of our everyday culture; here lies their specialness' (Clifford and King, 1987, Introduction). To help suggest this specialness, Clifford and King quote the painter Camille Pissarro, writing in 1893: 'Happy are those who see beauty in modest spots where others see nothing. Everything is beautiful, the whole secret lies in knowing how to interpret it' (Clifford and King, 1987, p. 251); and they quote William Blake's 'Auguries of Innocence' against the segregation of nature and for an everyday enchantment:

> To see the World in a grain of sand,
> And a Heaven in a wild flower,
> Hold Infinity in the palm of your hand,
> And Eternity in an hour.
> A robin redbreast in a cage
> Puts all Heaven in a rage.
>
> (Blake, quoted in Clifford and King, 1987, p. 131)

As Clifford and King's comment against 'experts' suggests, Common Ground's upholding of the everyday is also a raising of certain forms of the unofficial. *Holding Your Ground* is endowed with a Foreword by David Bellamy, who writes of his bulging post bag and in particular of a letter from a Welsh miner telling of the countryside of his youth; its wildflowers, birds, ponds, tiddlers etc.: 'Then there was a vibrant living unofficial countryside which still kept the black valley and the lives of its people green.' Now, the miner writes, he takes his grandchildren out into 'a sterile valley, sterile of work, sterile of wildlife, sterile of any hope' (Clifford and King, 1987, Foreword). The emphasis on the unofficial in Common Ground's work is at times as a complement to the official, but more often as a zone of opposition. In a Common Ground booklet, for example, Tom Greeves terms the 'official view of the locality' 'stuck' away from the parish in county archives and Planning Departments as, despite its situation, 'a mass of wonderful information' (Greeves, 1987a, pp. 4–5). Here the 'official' is presented not as in diametric opposition to the everyday local but simply as for the time being out of connection with it. Elsewhere though the 'official view' shows less potential for redemption. In the same booklet Greeves cries rallying that 'It is you who know your area intimately – in more detail than the officials in distant offices who have the power to change your surroundings' (Greeves, 1987a, p. 1), and chooses to quote an unspecified local voice of grievance: 'The planners never listen to us, do they?' (Greeves, 1987a, p. 4).

'Unofficial' knowledge, then, though it might function in a better future as an aid to 'official' decision-making, must for now it seems be defined as a form of knowledge distinct and distant from the detached and less than attentive official. Again Common Ground reverse the terms of an opposition, this time of expertise and lay knowledge, exhorting: 'Be positively parochial. Never forget that YOU are an expert in your place. No one knows what you think or feel unless YOU say so' (Common Ground, 1990b).

In their various publications Common Ground have given space to notably 'unofficial' landscapes of past and present, for example the plotlands of the inter-war years, places much vilified by the planning movement at the time (Ward, 1984; see Hardy and Ward, 1984; Matless, 1991b). An earlier unofficial landscape often cited is that found in the poetry of John Clare, a verse often claimed as radical, railing against enclosure (see Barrell, 1972; John Barrell is also a contributor to *Second Nature*). Clifford and King quote Clare in such a way as to emphasise his poetry of an unofficial, everyday landscape (Clifford and King, 1987, pp. 39, 83), and likewise Richard Mabey, who begins *The Common Ground* with a lengthy quotation from Clare's autobiography. Mabey comments: 'of all the descriptions and celebrations [of] . . . the English countryside's supposed golden past, it is the works of John Clare that we are turning to more and more for some kind of solace in its troubled present' (Mabey, 1980, p. 19). Clare, says Mabey, wrote of detail, of an agriculture 'of a piece with the natural world', and of loss. It is not simply Clare's subject-matter though but his manner: 'It is not just that what is under attack seems to be precisely the same now . . . what we recognize more is the sense of affront, of an invasion of personal territory by forces beyond our control.' Clare, for Mabey, not only portrayed a particular landscape invaded but a personalised landscape that was his 'not by virtue of ownership, but of familiarity . . . with losses that close to the heart it makes not a jot of difference if the cause is an enclosing landlord or a new motorway' (Mabey, 1980, p. 20).

Clare then is seen to present not merely an everyday and an unofficial but a personal landscape. As commented earlier, in Common Ground's work too the parish sits as a very personal place, a home. Place thereby becomes intimately bound up with a sense of self and of domesticity. Fraser Harrison presents the parish and its living landscape as a counter to the 'forces belittling our capacity for self-realization' (Harrison, 1986, pp. 131–2); Clifford and King term the parish with its 'Personal Landscapes' 'our home range' (Clifford

Figure 1.8 'Local Distinctiveness', cover of leaflet issued by Common Ground to support their Local Distinctiveness campaign, 1990, reproduced by permission of Common Ground.

and King, 1987, pp. 2–4), a personalised place, familiar and sustaining to the self. And in *Second Nature* Harrison ends his 'England, Home and Beauty' by asserting a particular version of domestic belonging: 'what must be conserved before anything else is the desire in ourselves for Home – for harmony, peace and love, for growth in nature and in our imaginative powers – because unless we keep this alive, we shall lose everything' (Harrison, 1984, p. 172).

On the front of their 'Local Distinctiveness' campaign leaflet (Figure 1.8)

Common Ground, below a drawing of a colliery wheel, make up a field, or perhaps a geology, in the form of a fingerprint; a local place stamped by, and apparent to, its belonging people.

b(iii). Unfixed settlement It is important to stress that in several senses Common Ground do not practise a simple extension of a 'rural descriptive tradition'. Description of place and landscape is at the heart of their work, but this is neither a consciously nor an unconsciously naive naturalism; the forms of representation they cite and employ are often not so straightforward. Description is also not merely for the sake of confirming what is there. Clifford and King, writing in *Holding Your Ground* on 'Celebrating Your Locality', state how the arts can 'help us to see familiar things in a new light', and how the practise of them can increase 'our powers of observation . . . enormously; we see shapes, outlines, details, colours which we had never noticed before or make connections we had missed' (Clifford and King, 1987, p. 184). The arts then bring not only descriptive reassurance but revelation, themselves casting difference on the place.

The artists Clifford and King cite in their chapter on celebration are sometimes those seldom found in a rural descriptive pantheon. Considering music, for example, Delius and Vaughan Williams are noted as expressing landscape and seascape, but also Olivier Messaien, a composer who while attending to nature and in particular bird-song might still be regarded by many as problematically 'modern' (Clifford and King, 1987, p. 202). Tom Greeves, writing on 'The Quality of a Place', makes a parallel point, in this case regarding a British aesthetic inertia when it comes to seeing beauty. Criticising the usual 'perverse' and pat dismissal of quarries as 'eyesores', Greeves suggests: 'We may be technologically developed as a country, but in terms of visual judgement and recognition of what is in a place we have probably never been so impoverished' (Greeves, 1989; Greeves' own eye for beauty of course balks at a landscape of modern consumption, a place where others might find a beauty). Common Ground's alphabet of local distinctiveness likewise embraces not only Charlotte Bronte, lapwings, pub sign and well-dressing, but also cooling towers, gasometer, gravel pit and wind (Common Ground, 1990a).

Common Ground's aesthetic of the local is not only though distinctive in its inclusion of features seldom before described as of value; it is also notable in not always conveying place as providing a solid or singular landscape of reassurance. This local personal is not one intended to simply and easily prop up the self. On a notable page of *The Living Landscape*, Fraser Harrison writes of how 'I have referred to our landscape as if it were a single, constant institution, like an oil painting, but of course that is not the case.' In his parish Harrison finds not one landscape but lots: 'Outside our windows there is not one landscape, there are myriads.' This multiplicity is for Harrison inherent in the visual, though not (at this stage of his argument) the physical world:

> The facts of geography may be named and numbered, but landscape is unnameable and unnumberable. Our physical world, despite the on-

> slaughts of modern farming, retains a long enduring consistency; our visual world, however, is merely a fleeting, quicksilver fantasy, a dream, a shadow playing on the cave wall, which dissolves and reforms, evaporates and stands again, approaches, draws back and sometimes slowly drifts, no more than a brown blur behind the hissing rain.
> (Harrison, 1986, p. 22)

This visual is a local world neither constant nor stable, a realm at times phantom, exciting yet at times disturbing, mind-bending for the observer:

> Our visible world is restlessly evanescent; it possesses less stability than a cloud running in shreds before a gale, less solidity than the surface of a brook tugged from below by its current and agitated above by the erratic breeze. Landscape was a mirage I never thought to question, but since coming here I have been disturbed and excited by its hallucinations, its continual bending and stretching, its ceaseless mutation of the physical out-of-doors on which I used so blithely to rely for constancy.
> (Harrison, 1986, p. 22)

This is a fluid, almost acid landscape. And it is only since living with a landscape, since being resident and familiar with a view, that Harrison has found that his settlement is not a place of stasis, not a place to settle at comfortable or complacent ease. Harrison's landscape then is less than solid, and there is also no one perspective on it. This is not a singular view, not a 'single, constant institution, like an oil painting'. Everyone will find a thousand different visual worlds. And for Harrison this multiplicity of perspective applies not only to an aesthetic dimension of landscape, but to the 'land' too. Harrison ends his first essay in *The Living Landscape* Hoskins-like, telling of the view from his window. His is not one view though, rather there are a number of perspectives open. Harrison's is less the performance of an historic guide than of a conjuror. Recalling from his childhood a street plan outside Lime Street Station in Liverpool where you could press different buttons to illuminate different features, he writes of how: 'By a simple trick of perception, I can produce a similar effect while standing at our window.' Harrison presses imaginary buttons to light up in the view a 'route map' of roads, an 'estate agent's plan' of labelled property, a 'parish topography' of locally named features, an 'agricultural study' of crop types and their value, a 'gamekeeper's plan' of game and vermin, and 'a sixth lights up the geography of our social life' (Harrison, 1986, p. 25). Different views then apply not only in the aesthetic realm. All of this shows to Harrison how 'our perception of land is no more stable than our perception of landscape'. While Hoskins brought you down to the local earth and its history after the trauma of 'The Landscape Today', Harrison by contrast won't let you settle:

> At first sight, it seems that land is the solid sand over which the mirage of landscape plays, yet it turns out that land too has its own evanescence. A closely observed, familiar and lived-in topography adds up to a very complex compound of signs and meanings, all of them in a state of perpetual change. Place – if place is the word to describe the unity of one's own land and landscape – is a restlessly changeable phenomenon, manifesting itself variously as 'parish', 'home', 'constituency', 'beauty

> spot', 'school catchment area', or even as 'prison', 'land of exile', 'suburb', 'commuter dormitory', and so on [the linking of these last four is telling]. The place we live in, look at, use, pay for, will not diminish itself to a simple, fixed title. Stowlangtoft is, for each of us, a separate and unique notion, as much a private symbol as a village.
>
> (Harrison, 1986, p. 26)

Not merely personal associations but cultural viewpoints make the place multiple, unfixed. Harrison's parish is less a rounded whole than a place of angles, of various perspectives and points of entry. In this connection it is interesting that Clifford and King cite as a possible pointer in their piece on 'Celebrating the Locality' David Hockney's practice of using collages of photographs of the 'same' scene from different angles in order 'to expand upon the single moment which is normally captured by just one photograph' (Clifford and King, 1987, p. 195). One such collage by Hockney appears in *Second Nature.*

It might seem that Harrison's emphasis on evanescence, on place unstable and multiple, contradicts with the themes of personal security and local continuity found elsewhere in Harrison and Common Ground's work. There seems a paradox here of two conceptions of place in time, one of continuity, the other of continual reformation. This may all be less paradoxical though if, as several contributors to Common Ground's work would seem to do, one works with a notion not of linear but of cyclical time. Again nature plays a key philosophical and rhetorical part. The cycle can suggest that a reference to visual evanescence or physical decay (the latter also frequent in Common Ground works) anticipates not disappearance but renewal. Indeed Richard Mabey ends *The Common Ground* by holding out just such a hope, arguing that 'the whole natural world is, in a way', a reminder 'that the alternative to progress is not stagnation but renewal'. This, writes Mabey, is the 'revelation' celebrated by Ted Hughes in the poem 'Swifts', with which *The Common Ground* ends:

> They've made it again,
> Which means the globe's still working, the Creation's
> Still waking refreshed, our summer's
> Still all to come-
> And here they are, here they are again. . . .
>
> (Hughes, quoted in Mabey, 1980, p. 251)

This is what you may see in a place, when you've seen it at least once before. So when in their 'Local Distinctiveness' leaflet Common Ground lament that 'nothing is left to decay gracefully', their's may not be a hope for a landscape of the tumbledown, but a call to allow a cycle to keep rolling, with grace; without decadence.

c. Working the local aesthetic: New Milestones and Parish Maps

Common Ground have worked their aesthetic in two notable projects, the 'New Milestones' sculpture project concerned to place art in landscape, and the Parish Maps Project concerned more with the representation of place. We end by considering the detail of each (attention is not given here to Common Ground's other main project, 'Trees, Woods and the Green Man', which is less directly concerned with issues of locality, focusing rather on the cultural, aesthetic and historical significance of trees and encouraging a contemporary celebration of them).

c(i). New Milestones The New Milestones Project, co-ordinated for Common Ground by Joanna Morland, aims in a sense to generate what Harrison termed 'parochial monuments' by encouraging and assisting local groups or individuals in commissioning sculptures to be set in the local landscape (Morland, 1988; Clifford and King, 1989). The Project was launched in 1985 in Dorset (partly as Angela King lived there and knew 'the land intimately' (Morland, 1988, p. 18), with the intention of expanding throughout the country afterwards (on subsequent developments see Morland, 1988, pp. 79–81). Four commissions had been completed by August 1988, and two others were in train. In keeping with other elements of Common Ground's work, the New Milestones, while acting as 'parochial monuments', are not intended to be monumental in size or dominating in effect: 'Rather than sculptures which dominate their surroundings, either in scale or presence, Common Ground is keen to encourage small-scale imaginative works which complement and conspire with the place' (Morland, 1988, p. 24).

The setting of a New Milestone is intended to heighten the work's humility: 'Less prominent locations are often preferable, where the work can be encountered casually or in thoughtful mood' (Morland, 1988, p. 24). This is not a sculpture to beg homage, rather one which conjures an intimacy between observer and object. Two of the sculptors involved in the Dorset work, Peter Randall-Page and Simon Thomas, express precisely this intention in the book detailing the project (Morland, 1988, pp. 29, 37). Common Ground's own emblem (Figure 1.9), a small stone engraved with the organisation's name and shown in publications idling in the grass, happened upon, itself expresses the aesthetic of a New Milestone (see for example Clifford and King, 1987, frontispiece).

The New Milestones are presented by Common Ground as very much a modern, contemporary art, yet one also with a primitive air. Here their setting in Dorset plays a part. Clifford and King write of how they were 'drawn to the cultural continuity' revealed in the apparent history and prehistory of the Dorset landscape (Morland, 1988, p. 18). This prehistory is highlighted too by Richard Cork in his 'Historical Perspective' in New Milestones (in which he places the sculptures in the modern yet primitive twentieth century British tradition of Gill, Epstein and Moore (Morland, 1988, pp. 7–9)): 'Reminders of Dorset's prehistoric identity abound; and since the country already boasts a remarkable array of stone crosses, obelisks and standing stones, the New Milestones Project centred on reviving this tradition in contemporary terms.' This is not for Cork though a revival of tradition against the contemporary:

Figure 1.9 'Common Ground', stone, lettercutting by Richard Grasby, Frontispiece of A. King and S. Clifford, *Holding Your Ground*, 1987 (revised edition), reproduced by permission of Common Ground.

'A proper awareness of the past does not, mercifully, allow degeneration into pseudo-archaic excursions. All the sculptors . . . have resisted the temptation to ape spectacular precedents like the White Horse at Weymouth. The new works belong firmly within their own time' (Morland, 1988, p. 10).

Both Thomas and Randall-Page allude to prehistory in their 'Milestones'. Thomas carves his work (Figure 1.10) as wooden grains of wheat lying

Figure 1.10 'Grains of Wheat', by Simon Thomas, West Lulworth, Dorset, 1987, part of the New Milestones Project, reproduced by permission of Common Ground.

between current grain fields and a grassland Site of Special Scientific Interest last cropped in the Bronze Age. Morland writes of how 'The grains would create a fertility bridge in space and time between the "Celtic fields" in the area, traces of which are still visible, and the grainfields of today' (Morland, 1988, p. 38). Randall-Page, on a site not far from Thomas' piece, makes his 'wayside carvings' (Figure 1.11) in fossil shape as 'a kind of tribute to the ancient lives which now constitute our terra firma'. The chalk land they sit on is composed of myriad small fossils; Randall-Page's carvings thus for him

Figure 1.11 'Wayside Carvings', by Peter Randall-Page, West Lulworth, Dorset, 1986, part of the New Milestones Project, reproduced by permission of Common Ground.

'strike up a resonance with the surrounding landscape by making a distillation of certain aspects of it' (quoted in Morland, 1988, p. 29). The carvings also make a medieval allusion, sited like wayside shrines in niches along a track (Morland, 1988, pp. 32–3).

While the Milestones then look to the past and primitive, they are yet contemporary, arching back to prehistory yet pointing forward. Joanna Morland writes of her hope that the sculptures 'will be as evocative of our times as milestones, stone crosses and other landmarks were for our predecessors' (Morland, 1988, p. 23), and Clifford and King, 'Putting the New Milestones Project in its Place', present it displaying a conviction 'that our moment in history has something to offer'. With its project Common Ground might they hope 'initiate new cultural touchstones worthy of our time' (Morland, 1988, p. 15).

The New Milestones are very much conceived as intimately local monuments, in their material, their making, their purpose and their placing. The sculptors use material – stone or wood in the Dorset works – found in the locality (Morland, 1988, pp. 32, 38, 43–4), and the making of a New Milestone is very much a local process, both in the sense of the sculptor working if possible on site (Morland, 1988, p. 22), and in the involvement of local people in the process. The making of the sculpture is reliant on local commission, whether by an individual or group (the majority of the Dorset commissions were from individual landowners sympathetic to the aims of the project). Beyond the role of the commissioner, though, Common Ground seeks to involve local people in consultations regarding the theme and site of the work:

> Artists are not encouraged to arrive with a ready-formed idea or to make a work without reference to the people or the place. Local people are a rich seam of knowledge, experience and feeling about their place – our aim is to weld this expertise with the creativity, vision and practical ability of the artist.
>
> (Morland, 1988, p. 22)

The intention with all the Milestones is to express and generate local meaning. Morland writes that the sculptures should 'be rooted in and speak of their place' (1988, p. 23), should be consciously local landmarks rather than be notable for their idiosyncracy. To make them still more local, none of the works was given a public name by the artist. There are no plaques to explain their presence. This was partly that 'passers-by might then see the work as surprise objects in the landscape, unfettered by notions of ART, and free of feelings of inadequacy or alienation' (Morland, 1988, p. 79), but also that the works might acquire their names locally. Morland notes with satisfaction how Randall-Page's wayside carvings have been found marked in a book of local walks as the 'three snails', and reports how they are 'becoming polished by the touch of many hands . . . and offerings of flowers have been found laid on top of the walling alcoves' (Morland, 1988, p. 35). And of another almost completed yet unnamed work, Morland is content to write: 'The work will undoubtedly acquire a name in time', and that when it does it will be cut into the stile which leads to it from the main track (Morland, 1988, p. 55).

Another of the New Milestones sculptors is Andy Goldsworthy, who made a large upright wooden ring of curved logs on each side of a road to mark an entrance to Hooke Park Working Woodland (Morland, 1988, pp. 40–7). Morland records how on a subsequent visit Goldsworthy was 'delighted . . . to be accosted by a lorry driver asking the way, who showed him a scribbled map with "The 2 Rings" shown as a waymark for finding his route' (Morland, 1988, p. 47). Goldsworthy has worked with Common Ground on a number of projects besides the Dorset rings. His work features prominently in *Second*

Nature (Mabey (ed.), 1984, pp. 80–2), and he has been closely involved in the 'Trees, Woods and the Green Man' project. Of the many well-known artists associated with Common Ground (Goldsworthy's work has an international reputation, and a large retrospective toured in 1990) it is Goldsworthy who has the closest tie, and his work has for a number of years exercised Common Ground's aesthetic. The hole motif found in the 'Two Rings' occurs frequently in his work, notably in a series of leaf works circled around central holes of black. Goldsworthy writes of the hole as a deliberately unsettling presence and absence: 'Looking into a deep hole unnerves me. My concept of stability is questioned and I am made aware of the potent energies within the earth. The black is that energy made visible' (Goldsworthy, 1985, p. 4). Goldsworthy's work in general emphasises instability and change, the kind of evanescence Harrison found in his landscape, the delicacy of what might appear permanent. His sculpture is rarely a monumental fixture, indeed often passes soon after making. The snow he uses melts, the leaf structures fall or blow away, beach works are washed by the tide. Fragility in nature is here embraced rather than being seen as a cause for worry. This is a 'fragile earth' which does not want bolstering: 'My sculpture can last for days or a few seconds – what is important for me is the experience of making. I leave all my work outside and often return to watch it decay' (Goldsworthy, 1985, p. 4). Goldsworthy lends his work some permanence by recording it in photographs, themselves monuments to a fleeting minute or two, lending 'a delicate hint of permanency' (Alberge, 1989). The sculptures themselves though do not persist. Even in the case of the 'permanent' commissioned two rings at Hooke Park, the work has been planted with holly bushes to soften the rings' junction with the road bank 'so that the work will become part of the place, not something separate, added and alien' (Morland, 1988, p. 46). Goldsworthy's are not monuments to proclaim themselves. In his general practice, then, as well as in his commissioned work, Goldsworthy expresses much of Common Ground's philosophy. Indeed like Common Ground he states that: 'In many ways my approach to the earth has been a reaction against the abuse of the land by the industrial farmer' (Goldsworthy, 1985, p. 5), and, again echoing Common Ground, spoke a phrase on the radio which could be a motto for his and their work: 'To find richness in what appear to be very humble places' (*Third Ear*, BBC Radio 3, 30 June 1989).

c(ii). Parish Maps Common Ground's philosophy of the local and familiar is embodied still more strongly in their Parish Maps Project (Greeves, 1987a; Clifford and King, 1987, pp. 287–9). There have been two strands to this, a number of maps commissioned from artists of places holding special significance for them, and a concurrent encouragement of local communities to produce maps of their parish.

The artists' Parish Maps were gathered in an exhibition. 'Knowing Your Place', in 1987, a display intended also to encourage community projects. The 'maps', covering places in city, town and countryside, clearly show Common Ground's intention that a Parish Map need not be bound by cartographic

Figure 1.12 'Common Ground: Regents Park', by Adrian Berg, 1987, Parish Map, reproduced by permission of Common Ground.

convention, alluding as much to the primitive maps often dismissed as 'works of the imagination' as to the scientific Euclidian procedures of the Ordnance Survey (see Avon Parish Maps Project, 1990, p. 1). Common Ground subsequently issued a number of the artists' maps as postcards. Some indeed display 'conventional' mapping and surveying techniques, whether in including an Ordnance Map in the work, as in Conrad Atkinson's map of Cleator Moor, where a background map is embellished with the artist's comments, comments which serve to distance Atkinson's 'emotional' and 'political' representation from the 'dry' plan view of the Ordnance Survey; or in adopting a composed prospect view, with the parish clearly laid out, as in Norman Ackroyd's 'memory picture' (quoted in Common Ground, 1987) of the parish of Swinbrook (on mapping and the prospect view see Daniels, 1990). Adrian Berg's 'map' of Regent's Park (Figure 1.12) similarly lays out the place, the prospect device here serving to make a contrast between the view of the park from the front of Berg's flat, which itself seems to obey the order demanded by the prospect, and the view to the rear, shown in the bottom section of the map, which in its (to the viewer presumed) modern confusion would seem to defy the prospect's demands of regularity (the thin horizontal strip in the middle of the map is a view of the terrace containing Berg's viewpoint over the park).

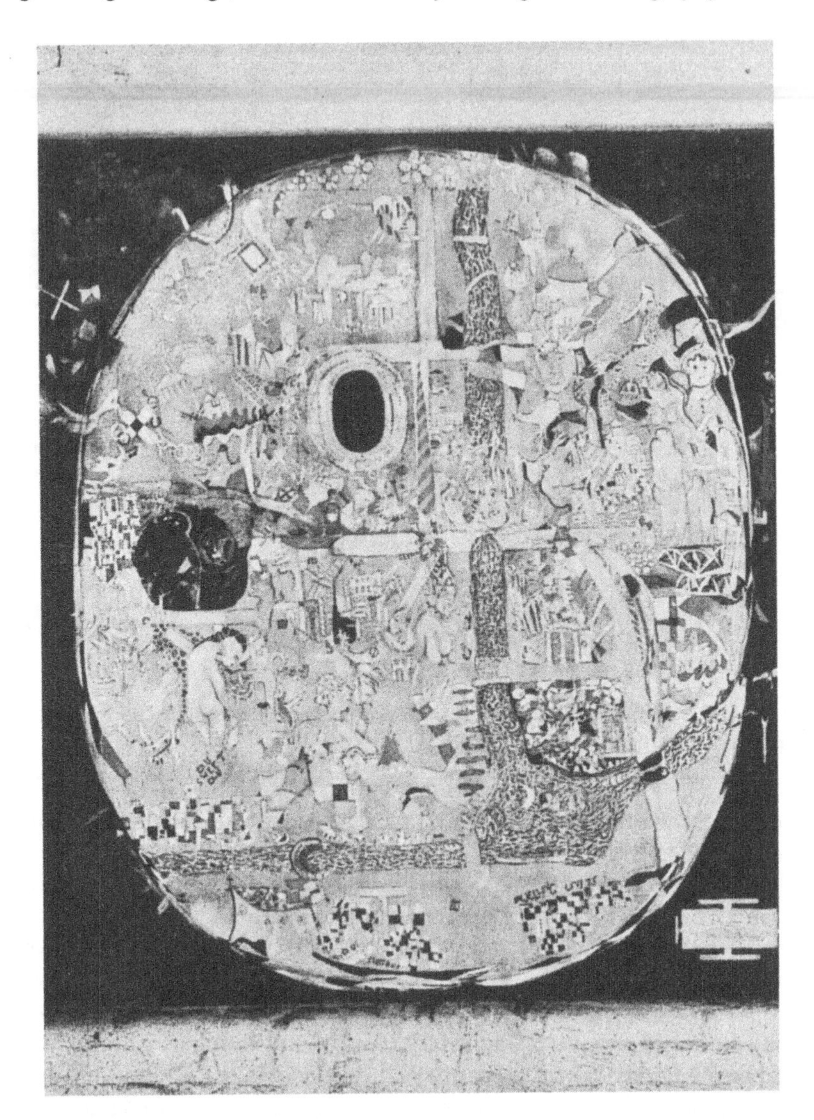

Figure 1.13 'The Real Centre of the Universe', by Balraj Khanna, 1987, Parish Map, reproduced by permission of Common Ground.

Other maps though abandon the map and the prospect as overall compositional devices, retaining a semblance of the Ordnance Survey only to saturate cartography with a self-consciously 'poetic' response to place, Thus Pat Johns' woven 'Topsham Observance' takes an aerial angle on the village's layout beside the Exe estuary and fills the landscape with a wood-spirit and a poem, and Tony Foster keeps a map almost as a locating insert beside pastoral images, soil and plant samples and a small idyll-detonating note that this is a test site for a nuclear power station. In the city, meanwhile, Stephen Willats'

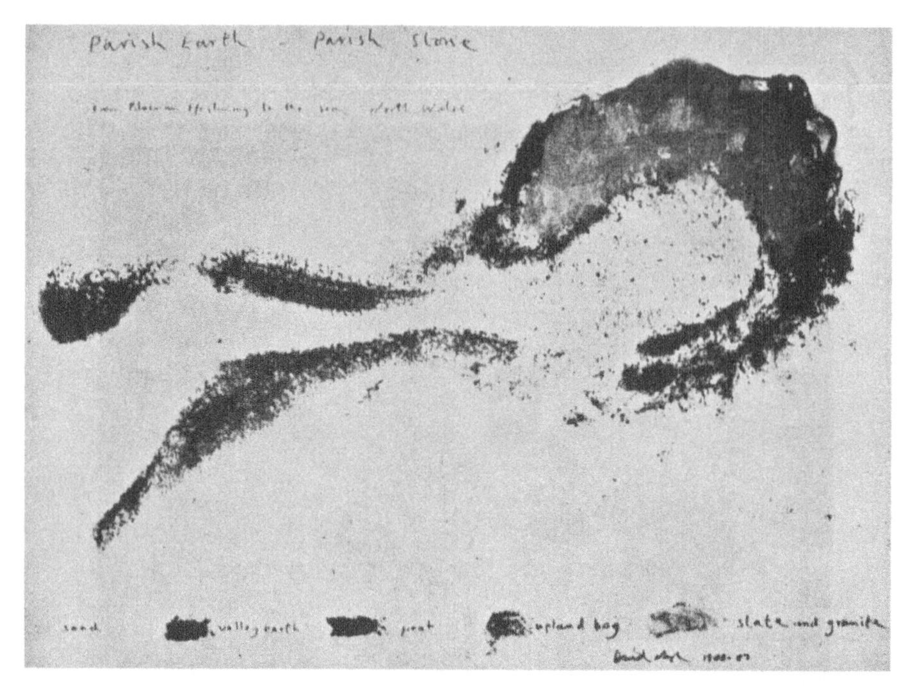

Figure 1.14 'Parish Earth, Parish Stone', by David Nash, reproduced by permission of Common Ground.

distance from the ground at the top of a tower block, a distance allowing him to vertically delineate the surroundings and the sites of found litter, serves to underline that his is a map of 'peoples' everyday social alienation' (quoted in Common Ground, 1987), of detachment from place.

Three other artists' maps also take the plan view, but not to express detachment. These maps, by Balraj Khanna, David Nash and Simon Lewty, seem to me to be those which best fulfil Common Ground's intention to both celebrate and shed a different light on the local and familiar. Khanna (Figure 1.13) makes Maida Vale in a map styled half as Mappa Mundi and half as Hindu mythological epic. Either way the teeming place, clearly edged and with a north-sign flying off to the top left, is 'The Real Centre of the Universe'. David Nash's 'Parish Earth, Parish Stone' (Figure 1.14), a map of the valley from Blaenau Ffestiniog to the sea is a simple construction made with the earth and stone found at the corresponding places in the catchment area. Nash brings the local land on to the paper to make up one aspect of the area. Nash is another whose work featured in *Second Nature* and who at one stage was to undertake a New Milestone for Common Ground (Morland, 1988, pp. 70–1). Simon Lewty's map of Old Milverton (Figure 1.15) is, like Khanna's, a denser and more complex piece, a map constructed as a journey which makes of the parish a place of stories. Naive representations of topography mix with words telling of the landscape and what can happen in it. A personalised geography emerges, Lewty's Old Milverton like no one and everyone elses.

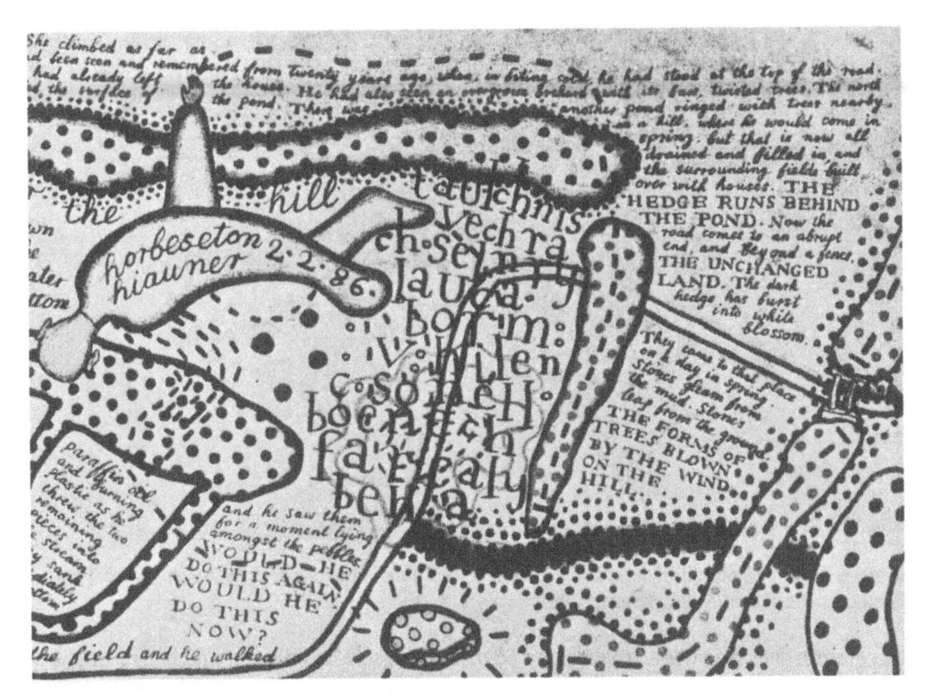

Figure 1.15 'Parish Map (Old Milverton)' (detail), by Simon Lewty, 1986, Parish Map, reproduced by permission of Common Ground.

All the Parish Maps, whether by individual artists or community groups, and whether they include writing or not, seek to articulate a visual language of place. There is a parallel in the effect sought in the arrangement of the *Second Nature* volume, a mix of written essays and pictorial works where, in Clifford and King's words, 'the visual works do not "illustrate" the essays, they are individual statements in another language' (Mabey (ed.), 1984, p. vii). The artists' maps were commissioned by Common Ground partly as examples of the kinds of visual articulation that community maps might follow. In the Parish Maps booklet issued to promote the project and guide prospective mappers Tom Greeves stressed the flexibility of convention and medium in a Parish Map: 'The map can be in any medium (paper, textile, ceramic, paint, photographic etc)', and suggested that the map need not even be pictorial but could embrace poetry, drama or music (Greeves, 1987a, p. 6). Most of these media have indeed featured in the maps made. The degree of departure from the conventions of cartography in the over 500 community maps produced so far though varies (as in the artists' maps) considerably, often simply because of the particular artistic skills and inclinations of the individuals involved in making the map. The maps in the county of Avon, for example, have varied from a ceramic map of multiple perspectives (Figure 1.16) and a map made of local plants to more straightforward arrangements of local photographs around a central piece of cartography.

More consistently self-consciously 'inventive' interpretations of mapping

Figure 1.16 'Windmill Hill ceramic map', by Jo Young, reproduced by Avon Parish Maps Projects, *The Parish Maps Project*, 1990.

have emerged from the placement of artists in schools. In Avon, for example, four artists have broadened the notion of local mapping well beyond the standard school recording of local land use (Avon Parish Maps Project, 1990).

Whatever the medium and techniques employed Common Ground intend the maps to express distinctively local knowledges and affections, and have a practical use in communicating that knowledge to others. Displayed in a prominent place, 'The map should become a focus of community pride and should help develop a greater sense of responsibility towards the total environment' (Greeves, 1987a, p. 6). The Parish Maps Project then illustrates well Common Ground's attention to the familiar and to the visual language of a place. It also though illustrates what seems at times a naive notion of 'community' as a unified and harmonious collection of individuals. There often seems little recognition, despite writings such as Harrison's on a place's multiplicity, that there will be more than one definition of place, and that these definitions will often be quite incompatible, unable to be embraced by one Parish Map. Common Ground though are by no means unique in expressing such a notion of community, and the Parish Map remains a fascinating document, perhaps less for any consensus picture that might emerge than for the tensions of viewpoint which are almost bound to surface in its making and its subsequent display. I remember attending an initial Parish Map meeting at which Tom Greeves spoke, invited after interest had been kindled by a local artist. All present were interested in what he had to say, and all tried for their say after. The parish council, who had the meeting under their auspices, had the first say. Local personalities began to bubble. Who would be on the map committee? Who could speak and draw for the place? Fascinating jockeying. Months later it was bogged down. I've no idea what happened in the end. Perhaps a map is on display on the library wall, as was mooted by some people, or perhaps nothing came out. Perhaps there should have been hundreds of maps done, one by each person in the village. There doesn't seem an easy way to express a community, because there's rarely one community to express.

One might also criticise Common Ground for a lack of attention to the potentially restrictive and confining sides of 'place' – the fixity of roots, the ties that bind, the lack for some of any way out, unhappy belonging made worse if you have few belongings; one should though I think echo them that operating at the local can be enabling, though what it enables varies mightily.

8. CONCLUSION: SOUNDING THE ENGLISH VILLAGE

The preceding six essays followed one another generally in chronology. They were not written though to convey the English village to the present day, to a point where what we have now is the village of Fraser Harrison, Common Ground, or whoever. The whole six essays persist to today, if with varying prominence and sway. And there are certainly elements of the English village imagination which have been missed here. A couple might be mentioned briefly; the village of the colour supplement, and the village of conservative ruralism. The former is a gentle, nigh on insipid country of the imagination expressed since the days before Sunday colour magazines but which has found a contemporary home there; the latter, also of long back catalogue, is an agrarian village of simultaneous social hierarchy and harmony, continuous down the ages and through the industrial revolution, and renewing now its long-time theme of a 'stewardship' rather than an 'exploitation' of the land. Both these villages and countries warrant sustained attention which cannot be given here. Echoes of them, though, indeed references to them will have been read in the preceding essays; these various imaginations do not operate in isolation. All these villages nourish each other in tension; replenishing, reacting, moulding and remaking.

By way of a conclusion here some nagging themes of the six written essays will be brought out; the sources of village difference, the geographical and formal particularity of 'the English village', the time of the village, the consumption of the village, and the theme of belonging. Finally the various ways of telling the village, and the authorities embodied in them, will be addressed.

i. The difference of the village

What then marks out the village in the English imagination? What features give it that distinction which urban and suburban people often desire to place themselves in? What makes the difference of the village? A number of themes can be brought out.

Village people

'The Country*man*'. The country*woman* tends to live in this imagination as a creature of the home, with the man as the visible public being, working out-of-doors, standing out for the country. More on village domesticity below. What though denotes this country*man*? There would seem to be a certain physique, even a genetic distinctiveness, picked out here and there by Bonham-Carter, and traced by Blythe in his *Akenfield.* Blythe indeed recalled it in 1990 as having been detectable 'until just the other day' (Blythe, 1990). There would also seem to be a difference in philosophy to some singular 'townsman'; this as well as landscape is held to make town and country distinct. And this philosophy is crucially connected by Bonham-Carter, by Turner, by Blythe, to land; to knowledge of it and connection to it, such that the country produces and contains a fundamentally different 'way of life' to town and city. This notion of a different way of life in the country is certainly still found today, though with a difference. The centrality of land's attendant countrymen has passed in the writings of a Harrison or a Mabey. Indeed as agriculture has become an enemy of nature, so, with the decline in the agricultural labour force which made up so many of 'the countrymen', an agrarian culture of nature is also seen to pass. There is here a highly complex and ongoing story of changing meanings and circumstances of agriculture, land and nature. One must always be attendant to just what a particular commentator means when using such seemingly given, 'natural' terms.

Now of course in many places the residents of the country and the 'country' men and women are not one and the same, if indeed they ever were entirely, although those working in agriculture are still often presented as the 'real' village, the 'authentic' people (Strathern, 1982). When Harrison in 1986 argued for his own status as a *villager*, living a different way of life to the town, and belonging to the place, it is significant that he did feel the need to *assert* this, and that it could not be taken as given. He did not though attempt to denote himself a '*countryman*'. One can claim membership of a place, but not so easily of something deemed almost a species.

Village order, village meaning

The village sits in the imagination in order; in visual order, of which more below, and in social order. The degree of stability, of regulation, of rhythm seen to pervade village life varies, and change is not denied in most accounts; in the village though change would seem to proceed without *disorder*. The imagined village offers an example and a hope of an ordered life, not in the sense of imagined discipline and systematisation but of stability, gentility and above all *harmony*, social and scenic. In this gentle rhythm and harmony the village seems an almost musical place, though of a particular kind of music. Disorder, dischord, is absent, located elsewhere, especially in the city. This village order/city disorder dichotomy is of course not always asserted; Thomas

Sharp for instance, like most planners of that period, upheld a rural order as different but complementary to an urban order. Others though – Turner in Luccombe, for example, and Harrison – present the urban as a contrary confusion and chaos. In this instance the order/disorder dichotomy becomes associated with another, that of *meaning* and the meaning*less*. Meaning is found in the order of the village, with urban life meaningless in its confusion.

The meaning of the village is in part environmental, of which below, and in part that of a particular form of social relations, that of community (see Harper, 1989). Certainly from the earlier accounts considered in this essay there emerges a village community formed very much of stable hierarchies in class and gender. A key theme in this imagined community, one related to the figure of the 'countryman' considered above, is an apparent absence of politics. Politics of an urban kind are 'out of place' in the village, and a boundary can be drawn between rural and urban concerns. And a cultural boundary can also be drawn around the village itself. Though the village is by no means cut off from other society, and indeed shares the trials of other villages, in this type of community (in contrast by implication to others) one *can* pursue things at a purely local scale, bounded. This tie of meaning and boundedness would seem to pervade the imagined English village, whether or not that village is presented as a place of hierarchical community, populated by 'countrymen'. It is a tie which still holds, for example, in Harrison's Stowlangtoft.

ii. A particular village

The boundedness of the village is of course linked to its position within open countryside. This most obvious 'difference' of the village, that it is a small settlement set in a landscape, is one often overlooked in discussions of what makes a village different. Not the least of reasons though why this obvious difference needs stating is that it is generally expressed in very particular terms. All villages may be by definition relatively small and in the country, but the village foregrounded in English imaginations tends to be a particular version; nucleated (see Wood (1991) for a similar American example), clearly bounded, with particular built features and a specific topographical setting. It is worth remarking here that although Sharp's village anatomy was offered in this essay as in many ways different to the other accounts considered, on the question of village form Sharp's is a like settlement. This nucleated English village tends to have a focus, often the church, physically, historically and emotionally at the core (one only has to visit a few villages to realise how particular a layout this is). And the place, clearly edged, seems to have its land, its parish, gathered around it. Often the village is presented grouped in a valley, nestling. Nature bounds the parish with a watershed. This placing is a recurrent twentieth century English topographical motif, in art, writing, map covers, Geography and more (Matless, 1990c). Village and landscape in visual symbiosis.

This placing of the village in its land, clearly bounded yet connected, of course serves to heighten the village's rurality, and conversely its non-urbanity.

Again this may seem an obvious point to make. A village, after all, is not a city, and the attraction of the village as a place non-urban is often emphasised. There is another seldom made point though about the visually marked rurality evident in the view of the nucleated village in the valley. The village is not only marked out as non-urban, it is also clearly *non-suburban.* One could read this non-suburbanity as simply an extended anti-urbanism, but one could also read it, with different implication, as appealing to a complementary dualism of town and country. That dualism, made clearly by Sharp, is a common one in twentieth century English landscape writing, planning and design, expressing not anti-urbanism but anti-suburbanism, with the suburb seen as visually and culturally blurring the polarity of city and country. Village, town and city are here presented as 'proper' settlements; the suburb is by contrast a place lacking an identity, a focus or a purpose. Given then this particular strand of the English culture of settlement it would perhaps be too simple to group the authors considered in this essay as anti-urban. One must of course make distinctions here between types of urbanity, but also between the country as an oppositional alternative or a complement to the city. The need to make this latter distinction was certainly evident when talking to people living in certain Cotswold villages in 1990; the form of settlement given as not only different but antagonistic to the village was less the city than the suburb. City and country remain a powerful double act in the English imagination; proper settlements, with values attached.

The English village is then a particular form of village, though one which we have seen in this essay to be often presented synecdochically as applying in general. One might also though argue that this village is not only formally but regionally particular. Donald Horne (1969), Martin Wiener (1981) and others have written of a 'Southern metaphor' at the heart of particular powerful notions of Englishness. In this metaphor the virtues of England – tradition, order, religion, idiosyncracy – are located in the South, and especially the rural South, and embodied in a matching symbolic scenery. Alun Howkins describes this as 'rolling and dotted with woodlands. Its hills are smooth and bare . . . it is cultivated and it is post-enclosure countryside' (Howkins, 1986, p. 64). I would argue that there is also a particular relation of landscape and settlement embodied in this metaphor. This is not only a rolling and cultivated landscape; it has nucleated villages as its foci. Turner's Luccombe, Blythe and Harrison's Suffolk, tap into this metaphor of a nucleated England. A topography of settlement is at a heart of this element of the moral geography of England.

It is perhaps though too easy to write in general of a Southern metaphor, with landscape and values in a snug fit (just as it is perhaps too easy to make of these values a singular conservative group). While we might find the Cotswolds, the Sussex Downland and parts of Suffolk being described again and again in such terms, other parts of 'the South' figure less prominently, and less obviously. And many authors who might themselves be thought of as fitting into the Southern metaphor's broad scheme of values tell of places other than the South. Hoskins is a good example. Hoskins may have evoked tradition, order and the idiosyncratic, and ended his *Making of the English Landscape* on the Oxfordshire dipslope of the Cotswolds, but he wrote at most length of the decidedly unSouthern East Midlands. Writing of the neglect of that area by the 'books by the score about the Cotswolds and the Chilterns . . .; the "Shakespeare country" . . . and East Anglia . . . ; the Pennines and the Peak', Hoskins located his Englishness firmly on East Midland ground: 'it is in some ways the most English of all the various provinces of this rich island'. Introducing his Batsford 'Face of Britain' volume on the region, Hoskins declared how he hoped to bring out 'something of this solid "Englishness" ' (Hoskins, 1949, p. v). There would seem to be more of geography, and especially more of regional geography (Daniels, 1991) in this metaphorical England than simple Southernness.

iii. The time of the village

What is this English village's place in time? Actually there are several, relating in different ways to past, present and future. Sharp's English village, for instance, sits on a continuum; Hoskins', by contrast, is a refuge against today and tomorrow. Different English villages look differently to progress, tradition, modernity, indeed embody different notions of what constitutes these three.

It seems often to be assumed that the English village lies on the side of tradition against modernity, with those two terms in opposition. This assumption certainly holds in the England of W. G. Hoskins, but many of the villages described in this essay do not fit into that simple dualism. For Sharp, for instance, tradition and the modern were not opposed. Likewise, though to a different end, Common Ground do not set up their parishes to be mapped as refuges from the modern world; indeed their readings of rural place engage with the arts of modernism. One could argue I think that these different times of the village are a significant element in the heterogeneity and complexity of the English village suggested in the Introduction to this essay. Patrick Wright has made similar points to these in his recent writings on conservation and Englishness, where he has criticised the way in which arguments have 'tended to reduce the whole post-war history of Britain to a polemicised opposition between the traditional nation and the various forms of modernisation that appear to threaten it' (Wright, 1990, p. 28). Of the accounts considered in this essay Hoskins' could certainly be characterised in this way, indeed one could argue for Hoskins as one of the founders of the polemic, but the point is that his village, his England, is just one, though a very potent one, among many. Wright too signals such a plurality when he argues against characterising 'the themes which now cluster around the landscape – questions of place, cultural particularity, conservation and expressive quality' as necessarily narrowly parochial, escapist and anti-modern (Wright, 1990/1, p. 7; see also Matless, 1990a, 1990b). For Wright these themes of landscape are complex, fascinating, unsettling, even strange: 'recent English culture – *even* recent English culture – begins to seen positively exotic: full of barely imaginable possibilities', possibilities which 'simply don't fit' into any assumed polarity of tradition and modernity (Wright, 1990/1, p. 7). This essay has taken themes of English settlement and landscape in like vein.

There are other times of the village though, times which evade altogether debates on the continuity or antagonism of past, present and future. These place the village in non-linear time, or outside any time at all. Both of these, in different ways, refuse the opposition of tradition and modernity (though that should not be taken as necessarily an argument 'for' them). In several of the accounts considered in this essay – Blythe's and Mabey's, for instance – the village, because of its connection to agriculture and/or nature, is placed in cyclical rather than linear time. Village life is presented as in whole or part bound to the seasons. In such a time past and future bend round to meet, and it makes no sense to make an opposition of them. So Mabey ends *The Common Ground* declaring how 'the alternative to progress is not stagnation but *renewal*' (Mabey, 1980, p. 251). Other villages besides though miss questions of tradition and progress by finding a place *outside* time, in the eternal. There is a recurrent mystical theme in the English village imagination, making of the village a spiritual place (whether this is expressed in doctrinal terms or no) where time is no matter. Here we find, for example, the early nineteenth century artist Samuel Palmer, Ronald Blythe's Akenfield 'Poet', the late Peter Fuller and Fraser Harrison in Stowlangtoft church with their humanistic religious ethic, and an early twentieth century writer like T. F. Powys, in whose 1927 novel *Mr Weston's Good Wine* time stops for an evening in the village of Folly Down when Mr Weston, a figure of God, comes to purvey his wine to the residents (Powys, 1976). This is often a village of allegory, at times one possessed of the surreal; outside time with everyday life in the country transfigured.

iv. The consumption of place

Themes emerge as to how people consume the village; how and why they take it in, and what good it does them.

Different groups are presented as consuming place in different ways; insiders and outsiders, upper, middle and working class. Blythe, with his 1969 'national village cult', presents an early explicit analysis of middle-class incoming taste. It is I think important though to suggest that themes of 'consumption' should not be considered as applying only to incomers, or to the middle class. To read some writings on the idea of the English village one would think that it is only these groups who actually engage imaginatively with their place, while others simply live there. One could though argue for all kinds of people taking particular pleasures and nourishment from their village surroundings, engaging with them aesthetically, emotionally, imaginatively. 'Nourishment' is an apt word to employ here, for the village is invariably presented as a setting good for you, morally and physically, a place of healthy air and beneficient proximity to nature. This is a place to build both a body and a self (on links between notions of health, morality, English landscape and diet see Bishop (1990b)). The theme of the village and the self is returned to below.

This 'nourishment' of the village is, however, rarely phrased in terms of consumption. One might argue that this is simply due to that word not being particularly fashionable in rural commentary, but there is a story in such a fashion. 'Consumption' and 'landscape', 'consumption' and 'the rural', 'consumption' and 'the village', have long been terms in tension in English culture (Wiener, 1981). Consumption has often been lent a hint of immorality, of indulgence, of decadence, at odds with a moral discourse of health and vigour often located in the country. This tension can be linked into other recurrent English themes; cultural tensions between landed and industrial capital (though the two have seldom been entirely unconnected economically), and the notion of the countryside as a site of non-commercial values, a place where happiness comes from the non-material (an argument often linked to a notion of the happy rural poor). Whichever of these or other themes it might be bound up with, notions of the (im)morality of consumption remain strong in the discourse of the English village. And this moral discourse is of the political Left as well as the Right. Fraser Harrison, when he asserts the subversive eye over the voyeuristic, taps into this morality of landscape.

v. Belonging in the village

There are a number of senses in which the village is presented as a possibly self-making place, a number of reasons why it's thought better to belong there than elsewhere.

One is the village as a place of the ordinary. In a sense the whole imagined English village is an announcement of the ordinary, or, in architectural terms, the vernacular. Common Ground of course seek openly to assert the everyday, the common, but one could regard all the many imagined English villages as making a similar announcement. The village is presented as a place where the ordinary pertains in an intimate manner, a place contrasting with the supposedly anonymous and standard suburb and city. This is in large part an argument about 'alienation'.

The village is also told as a place where uniqueness, quirkiness, even eccentricity pertains, despite wider forces seeking its erasure (see Samuel (1989) on these themes and Englishness). The village is thus a reservoir of particularity, often set up as a counter to the values of commerce and the official state. The latter are presented as forces for standardisation, imposing uniformity over places. The village is given as expressing values of variety, of the individual, of the unofficial, of the particular. The village has contrived to remain a unique place, and the message is that its uniqueness can permeate you if you can belong there, with its and your identity mingling. How you can come to belong there, rather than simply reside, is, however, another matter, an issue, as Harrison's arguments for his own belonging illustrate, which can be at the heart of many contemporary disputes over rural place.

The question of belonging indeed crops up frequently in the imagined English village, in all kinds of accounts (for an anthropological angle on this see Cohen (ed.), 1982). The village is often put up as a personal place, in terms of community and domesticity. Villages are presented as communities, as places one publicly shares. They are also though public places which can become personalised. The whole village and parish carries personal meaning for the resident, meaning proceeding from familiarity and belonging. There is again perhaps a key distinction here between the imagined village and the imagined suburb. In the latter the personalised landscape seems to be that personally owned; home and garden, but not beyond. In the suburb property is home, in the village home is the whole place, and not least because of scale and boundedness. These most obvious definitional features again play a key fictive role.

The village is also often cast as a site of domestic belonging. There is a long binding in English culture between visions of village community and particular notions of domesticity, with the two combining to place the woman ideally in the home. Davidoff, L'Esperance and Newby have argued that from the early nineteenth century ideals of both home and community were commonly located in the village, and that these ideals were not simply sharing a location but were interconnected with it and each other: 'The very core of the ideal was home in a village community' (Davidoff *et al.*, 1976, p. 140). This ideal, this moral geography, embodied a very particular domestic as well as communal order, a hierarchy of gender as well as class. Davidoff, L'Esperance and Newby argue for the persistence of this moral geography to at least the mid-twentieth century, and one can certainly read a good example in Turner's Luccombe. And although it is less easy to trace as an explicit ideal in, say, Blythe's Akenfield or Hoskins' English landscape, the ideal is certainly not challenged, and its questions are not raised, such that one could argue that it maintains its cultural presence and power implicitly. Certainly it is hard to imagine anything other than that particular domestic order in Akenfield. And calling, as Fraser Harrison does, for attention to 'great themes of existence – birth, marriage, morality . . .' in a village account can invoke such a domestic order, whether the author intends it or no. The parallel of the communal and domestic ideals of the village, a parallel perhaps accentuated by seeing the village as a personalised place, lingers on and might well be conjured unawares in doing the English village.

vi. Telling the place

There are many ways of knowing this settlement, many forms of authority practised in doing the English village. Here I will pick out three; a plan-view, excavation and absorption.

Thomas Sharp took the plan-view, placed himself in the air as an expert.

The expression of expertise through an aerial overview was nothing new in 1946, indeed it had been a key motif in the inter-war work of Sharp and other planner-preservationists (Matless, 1990b); Sharp's *Anatomy* continued the view into the post-war country. What changed in the post-war period though was that another way into the English village emerged as a direct counter to the plan.

Sharp's village was a place of shape and surface, apprehensible and plannable to one with the right knowledge; other villages, other knowledges though are expressed in Turner's Luccombe, and Hoskins' landscape. Hoskins' village is one to be dug. His is an archaeological knowledge, the landscape being read as a document of depth, where meaning is under the surface, metaphorically and literally. While Sharp's is a planar village, Hoskins' is a place of deep reverie:

> For my own part I am not much interested in surface impressions. The three visible dimensions of . . . a landscape are not enough: they may entrance for the moment but they make no abiding impression on the mind. One needs the fourth dimension of time to give depth to the scene . . . there are depths beyond depths in the simplest scene.
>
> (Hoskins, 1949, pp. v–vi)

One detects here a suspicion, almost a distrust of 'entrancing' surface images, lending Hoskins' excavation a moral intent. Hoskins' presents his as a landscape of more weight, and of rougher texture. The metaphor of texture would seem to be an important one in relation to doing the English village (Daniels, 1989, p. 210), in particular in the form of authority it helps lend to the author. Texture tends in this usage to convey a communality to the landscape, to heighten the sense of it being a thing woven by many hands, the product of a long and various endeavour. This is not to imply that not using a metaphor of texture means one must be ignoring such communality; what the use of that metaphor can do though is to ascribe a certain humility to the contemporary author. In emphasising longevity, the gradual and the complex, the metaphor suggests a distaste for simple and dramatic replanning, and a patience in the author, a desire to be but one among many contributors to the landscape and the reading of it. Certainly Hoskins projects a more humble authority than, say, Sharp's assertive vision, and is perhaps more powerful for that humility.

Hoskins then presents a place to be excavated, and that theme of excavation continues to be a powerful one. Much of Common Ground's work, for example, feeds on this version of geographical history. Common Ground's work also though feeds off another way of knowing the English village, that of absorption, of the imbibing of a place through sustained contact. This is a way of knowing which prides residence rather than the occasional visit. Sharp's was the view of a traveller-planner, Hoskins, though emphasising settlement and deep enquiry, nevertheless offered a wide survey. Turner's Mass-Observed Luccombe, though, and Harrison's Stowlangtoft, are places known by staying there, places less to survey or patiently read than to absorb. Notions, and they are often no more defined than that, of empathy, insight and essence of place are brought to bear (there is an interesting parallel here with much humanistic geography). Often, it is true, these are seen as complementing survey and excavation, but they bring a very different kind of knowledge, and one often deemed superior. This English village is a place to hear, touch, taste and breathe. Often it is a place which by definition you cannot quite grasp, a place with a mystery at the heart, defying definition, defying regulation, defying a Sharp or a Bonham-Carter.

There are then a number of ways into the village. These are, however, by no means mutually exclusive, and many of the approaches considered in the essay draw on more than one. While Sharp's, for example, is a singular pespective, Common Ground's and Mass-Observation's are not single or simple knowledges. There is more than one way into Luccombe, for instance; through smell, through statistics, through conversation, through photography. Common Ground and Mass-Observation present plural villages, different versions being produced by different enquiries, several pictures lying alongside one another to make up a whole.

So there are many ways into the English village, and many combinations to be made to them. I am of course though still writing here as if 'the village' were sitting there as a physical object waiting to be 'done'. There is then a final point to be made here which muddles things further. This essay has argued that the village is a culturally charged object; charged with powers, knowledges, memories, politics, moralities and more. So in doing the English village you are 'approaching' something already speaking; meaning cannot help but be conjured beyond your intention. Here, as anywhere, description cannot purport to naturalism. You are stepping in a discourse where your 'object' is not some passive thing, indeed its cultural energy is such that the division of subject and object gets put into doubt. And of course you, as 'subject', relate to the English village in your own particular way, depending who you are. With all this meaning abroad, the village, as a 'heritage', if you feel it as such, cannot perhaps but help unsettle you, the enquirer, Foucault's 'fragile inheritor' (Foucault, 1986, p. 82). The settlement becomes a place of unsettlement; often, it is true, a gentle unsettlement but an unsettlement nonetheless, with various meaning resounding around. Peter Bishop, picking on an acoustic metaphor at the end of his study of imagined Tibet, comments how 'places echo' (Bishop, 1990a, p. 249). This essay too has sought to present an acoustic of an imagined place. 'The English Village' certainly echoes, and if nothing else this essay should have broached that whether writing, drawing, dreaming or recording it, one should attend to the soundings.

ACKNOWLEDGEMENTS

Thanks to Nigel Thrift for commenting on an earlier draft. Much of the research for this essay was carried out while working on a University of Bristol/ St David's University College Lampeter ESRC sponsored project on the impact of the middle class in the contemporary English and Welsh countryside.

Figs 1.4, 1.5 and 1.6 are used by permission of John Hinde.

REFERENCES

Alberge, D. (1989) Making an impression with the elements, *Independent*, 18 February, p. 43.
Avon Parish Maps Project (1990) *The Parish Maps Project: Know Your Place in Avon*, Avon Parish Maps Project, Bristol.
Barr, C. (1977) *Ealing Studios*, Cameron & Tayleur/David & Charles, London.
Barrell, J. (1972) *The Idea of Landscape and the Sense of Place, 1730–1840: An Approach to the Poetry of John Clare*, Cambridge University Press.
Berger, J. (1979) *Pig Earth*, Writers & Readers, London.
Berger, J. (1985) *The Sense of Sight*, Pantheon, New York (published in the UK as *The White Bird*).
Berger. J. (1986) Credibility and mystery, *Marxism Today*, Vol. 30, no. 10, pp. 45–7.
Bishop, P. (1990a) *The Myth of Shangri-La: Tibet, Travel Writing and the Western Creation of Sacred Landscape*, Athlone, London.
Bishop, P. (1990b) *Consuming Constable: Diet, Utopian Landscape and National Identity*, Nottingham University Department of Geography (Working Paper 5).
Blythe, R. (1969) *Akenfield*, Allen Lane, London.
Blythe, R. (ed.) (1981) *Places: An Anthology of Britain*, Oxford University Press.
Blythe, R. (1990) Constable Country, *Independent Magazine*, 25 August, pp. 40–1.
Bonham-Carter, V. (1952) *The English Village*, Penguin, Harmondsworth.
Cavaliero, G. (1977) *The Rural Tradition in the English Novel 1900–1939*, Macmillan, London.
Chamberlain, M. (1983) *Fenwomen: A Portrait of Women in an English Village*, Routledge & Kegan Paul, London (first published 1975).
Chaney, D. and Pickering, M. (1986) Authorship in documentary: sociology as an art form in mass observation, in J. Corner (ed.) *Documentary and the Mass Media*, Edward Arnold, London, pp. 29–44.
Clifford, J. (1988) *The Predicament of Culture: Twentieth Century Ethnography, Literature and Art*, Harvard University Press, London.
Clifford, S. and King, A. (1987) *Holding Your Ground: An Action Guide to Local Conservation*, Wildwood House, Aldershot.
Clifford, S. and King, A. (1989) New Milestones: sculpture, community and the land, *The Green Book* Vol. 3, no. 1, pp. 40–4.
Cohen, A. (ed.) (1982) *Belonging: Identity and Social Organisation in British Rural Cultures*, Manchester University Press.
Common Ground (1987) *Knowing Your Place: An Exhibition of Artists' Parish Maps*, Common Ground, London.
Common Ground (1990a) *Local Distinctiveness*, Common Ground, London.
Common Ground (1990b) Mayday! Mayday! (advertisement), *Independent*, 5 May, p. 39.
Cormack, P. (1978) *Heritage in Danger*, Quartet, London.
Daniels, S. (1989) Marxism, culture, and the duplicity of landscape, in R. Peet and N. Thrift (eds.) *New Models in Geography*, Vol. 2, Unwin Hyman, London. pp. 196–220.

Daniels, S. (1990) Goodly prospects: English estate portraiture, 1670–1730, in N. Alfrey and S. Daniels (eds.) *Mapping The Landscape*, Castle Museum/University Art Gallery, Nottingham, pp. 9–12.
Daniels, S. (1991) Envisioning England, *Journal of Historical Geography*, Vol. 17, no. 1.
Davidoff, L., L'Esperance, J. and Newby, H. (1976) Landscape with figures: home and community in English society, in J. Mitchell and A. Oakley (eds.) *The Rights and Wrongs of Women*, Penguin, Harmondsworth, pp. 139–75.
Deakin, R. (1989) The riches we ignore, *Financial Times*, 9 December, Weekend, p. xvii.
Dreyfus, H. and Rabinow, P. (1982) *Michel Foucault*, Harvester, Brighton.
Foucault, M. (1986) Nietzsche, genealogy, history, in P. Rabinow (ed.) *The Foucault Reader*, Penguin, Harmondsworth, pp. 76–100.
Fuller, P. (1985) *Images of God: The Consolations of Lost Illusions*, Chatto & Windus, London.
Fuller, P. (1988) *Theoria: Art, and the Absence of Grace*, Chatto & Windus, London.
Goldsworthy, A. (1985) *Rain sun snow hail mist calm*, The Henry Moore Centre/Northern Centre for Contemporary Art, Leeds/Sunderland.
Greeves, T. (1987a) *Parish Maps*, Common Ground, London.
Greeves, T. (1987b) *The Parish Boundary*, Common Ground, London.
Greeves, T. (1989) The quality of a place, *Countryside Commission News*, no. 38, July/August, p. 3.
Hardy, D. and Ward, C. (1984) *Arcadia For All: The Legacy of a Makeshift Landscape*, Mansell, London.
Harper, S. (1989) The British rural community: an overview of perspectives, *Journal of Rural Studies*, Vol. 5, no. 2, pp. 161–84.
Harrison, F. (1984) England, home and beauty, in R. Mabey (ed.) *Second Nature*, Jonathan Cape, London, pp. 162–72.
Harrison, F. (1986) *The Living Landscape*, Pluto, London.
Hewison, R. (1987) *The Heritage Industry*, Methuen, London.
Horne, D. (1969) *God is an Englishman*, Sydney.
Hoskins, W. G. (1949) *Midland England*, Batsford, London.
Hoskins, W. G. (1955) *The Making of the English Landscape*, Hodder & Stoughton, London.
Hoskins, W. G. (1959) *Local History in England*, Longman, London.
Howkins, A. (1986) The discovery of rural England, in R. Colls and P. Dodd (eds.) *Englishness: Politics and Culture 1880–1920*, Croom Helm, London, pp. 62–88.
Keith, W. J. (1975) *The Rural Tradition*, Harvester, Brighton.
Mabey, R. (1980) *The Common Ground*, Hutchinson, London.
Mabey, R. (1984) Introduction: entitled to a view? in R. Mabey (ed.) *Second Nature*, Jonathan Cape, London, pp. ix–xix.
Mabey, R. (ed.) (1984) *Second Nature*, Jonathan Cape, London.
Mass Observation (1987) *The Pub and The People*, The Cresset Library, London.
Matless, D. (1989) Mass Observation: review, *Journal of Historical Geography*, Vol. 15, no. 2, pp. 213–14.
Matless, D. (1990a) Ages of English design: preservation, modernism and tales

of their history, 1926–39, *Journal of Design History*, Vol. 3, no. 4, pp. 203–12.

Matless, D. (1990b) Definitions of England, 1928–89: preservation, modernism and the nature of the nation, *Built Environment*, Vol. 16, no. 3, pp. 179–91.

Matless, D. (1990c) The English outlook: a mapping of leisure, 1918–39, in N. Alfrey and S. Daniels (eds.) *Mapping The Landscape*, Castle Museum/University Art Gallery, Nottingham, pp. 28–32.

Matless, D. (1991) Ordering the land: the 'preservation' of the English countryside, 1918–39. Unpublished Ph. D. thesis, University of Nottingham.

Matless, D. (1992) An occasion for geography: landscape, representation and Foucault's corpus, *Environment and Planning D: Society and Space* Vol. 10, no. 1, pp. 41–56.

Matless, D. (1993) One man's England: W. G. Hoskins and the English culture of landscape, *Rural History*, Vol. 4, no. 2, pp. 187–207.

McLaughlin, B. (1983) *Country Crisis: The lid off the Chocolate Box*, Channel 4 Television, London.

Meinig, D. W. (1979) Reading the landscape: an appreciation of W. G. Hoskins and J. B. Jackson, in D. W. Meinig (ed.) *The Interpretation of Ordinary Landscapes*, Oxford University Press, pp. 195–244.

Morland, J. (1988) *New Milestones: Sculpture, Community and the Land*, Common Ground, London.

Newby, H. (1977) *The Deferential Worker*, Allen Lane, London.

Newby, H. (1980) *Green and Pleasant Land*? Penguin, Harmondsworth.

Philp. M. (1985) Michel Foucault, in Q. Skinner (ed.) *The Return of Grand Theory in the Human Sciences*, Cambridge University Press, pp. 65–81.

Potts, A. (1989) 'Constable Country' between the wars, in R. Samuel (ed.) *Patriotism*, Vol. 3, Routledge, London, pp. 160–86.

Powys, T. F. (1976) *Mr Weston's Good Wine*, Penguin, Harmondsworth.

Riden, P. (1983) *Local History*, Batsford, London.

Samuel, R. (1989) Introduction: exciting to be English, in R. Samuel (ed.) *Patriotism*, Vol. 1, Routledge, London, pp. xviii–lxvii.

Sharp, T. (1946) *The Anatomy of the Village*, Penguin, Harmondsworth.

Stansfield, K. (1981) Thomas Sharp 1901–1978, in G. Cherry (ed.) *Pioneers in British Planning*, The Architectural Press, London, pp. 150–76.

Stewart, C. (1948) *The Village Surveyed*, Edward Arnold, London.

Strathern, M. (1982) The village as an idea: constructs of village-ness in Elmdon, Essex, in A. Cohen (ed.) *Belonging*, Manchester University Press, pp. 247–77.

Summerfield, P. (1985) Mass-Observation: social research of social movement? *Journal of Contemporary History*, Vol. 20, pp. 439–52.

Thomas, K. (1983) *Man and the Natural World: Changing Attitudes in England 1500*–1800, Penguin, Harmandsworth.

Thrift, N. (1986) Little games and big stories, in K. Hoggart and E. Kofman (eds.) *Politics, Geography and Social Stratification*, Croom Helm, London.

Turner, W. J. (1947) *Exmoor Village*, George Harrap, London.

Ward, C. (1984) A place in the country, in R. Mabey (ed.) *Second Nature*, Jonathan Cape, London, pp. 198–206.

Wiener, M. (1981) *English Culture and the Decline of the Industrial Spirit, 1850–1980*, Cambridge University Press.
Williams, R. (1975) *The Country and the City*, Paladin, London.
Williams, R. (1984) Between country and city, in R. Mabey (ed.) *Second Nature*, Jonathan Cape, London, pp. 209–19.
Wood, J. S. (1991) 'Build, therefore, your own world': the New England village as settlement ideal, *Annals of the Association of American Geographers*, Vol. 81, no. 1, pp. 32–50.
Wright, P. (1985) *On Living in an Old Country*, Verso, London.
Wright, P. (1990) Revival or ruin? The 'Recording Britain' scheme fifty years after, in D. Mellor, G. Saunders and P. Wright (eds.) *Recording Britain*, David & Charles, London, pp. 25–36.
Wright, P. (1990/1) Englishness: the romance of the oubliette, *Modern Painters*, Winter, pp. 6–7.

Chapter 2

Habermas, Rural Studies and Critical Social Theory

Martin Phillips

CRITICAL RURAL STUDIES: AN INTRODUCTORY DISCUSSION

The notion of rural studies as a form of critical social theory has been advanced since the late 1970s (e.g. Newby, 1977; Newby and Buttel, 1980; Hoggart and Buller, 1987; Cloke, 1989). The earliest explicit calls for such an approach to rural studies were advanced by sociologists such as Howard Newby, but gradually the term 'critical theory' has worked its way into the rural literature produced by geographers, planners and environmental scientists (see Lowe and Bodiguel, 1990, pp. 44–6). Today a 'critical approach' is often contrasted against 'traditional', 'positivist' and 'empiricist' approaches (see Urry, 1984; Cloke, 1989; Robinson, 1990). Yet despite its common currency in the rural literature, there appear to be at least three problematic issues which as yet have been insufficiently addressed.

First, the notion of critical social theory is rarely spelled out in the literature of rural studies – a feature which incidentally is also common within some of its constituent disciplines such as geography (Phillips, 1994). This feature is made particularly problematic because there are notable differences in the way the term critical social theory has been used, some of which relate to distinct disciplinary contexts. In the case of rural sociology the adjective 'critical' was added by Buttel and Newby to imply 'a more independent and sceptical attitude towards . . . rural phenomena' and to prevent the rural sociologist being 'lulled into the condition which anthropologists refer to as "capture" – adopting a wholly uncritical stance towards the structure and institutions of rural society and thus rarely going "beyond the cliched and subjective experiences of the people he [*sic*] is studying"' (Newby and Buttel, 1980, p. 2). In other words, a 'critical rural sociology' was politically charged: it was prepared to reject the accepted structure, institutions and perceptions of the countryside. It was, in the words of Newby and Buttel (1980, p. 2), a 'critique of the status quo' in rural society.

Within geography the term 'critical' has taken on a slightly different inflection: rather than being a rejection of the status quo in rural society it has been used to signal a rejection of notions of distinctively rural spaces (see Cloke, 1989; Hoggart, 1988, 1990; Rees, 1984; Robinson, 1990; Urry, 1984). Critical social theory in this context has been taken to refer to theory which emphasised 'general' processes, particularly those associated with capital restructuring. As Cloke notes, such 'critical theories' as those of contemporary neo-Marxist political economy pose rather a conundrum for rural geographers: 'Accept the arguments of most political-economic theorists and the legitimation for *rural* categories of study largely disappears. . . . Reject these arguments, and the potential explanatory power of the political economy is removed' (Cloke, 1989, p. 175). How this conundrum is to be resolved is, as yet, far from clear (see Cloke and Davies, 1992; Halfacree, 1993).

What is much clearer, however, is that critical social theory – both in the sense of a critique of existing society and as a critique of a parochial form of rural studies – is regarded with some hostility within rural studies, particularly within rural geography. Robinson, for example, writes:

> Although in the past two decades rural geographers have focused more closely on specific problems, e.g. depopulation, declining services . . . the influence of urban sprawl . . . conflicts between recreation and conservation, they have mirrored human geographers as a whole by 'branching into anarchy' in terms of their kaleidoscope of foci, methodologies and general ethos. . . . Within this 'anarchy' two divergent views of 'rural' have emerged. One might be termed 'traditional'. . . . This approach has tended to eschew theory and closer links with developments within other social science disciplines. However, it has had quite a strong 'applied' component. . . . In contrast, a strong line of argument in the 1980s has combined a broader interdisciplinary approach with a philosophy that has regarded 'rural' as something of a false theoretical category. . . . This approach mirrors human geography in its faltering attempts to develop a philosophical base that will allow it to embrace a suite of methodological possibilities incorporating humanistic and empirical elements as well as the more restrictive, and generally more utopian, concepts within structuralism. It is possible to view the apparent chaos of the breadth of scope and plethora of methodologies in rural geography with some alarm.
> (Robinson, 1990, pp. 19–20)

The tone of this description of the developments of the 1980s is almost completely negative and it is clear that notions of 'critical social theory', of whatever form, are far from accepted within rural studies.

Cloke (1989) has tried to account for such reactions to the advocacy of critical social theory in terms of: (i) a desire to preserve the 'rural' as a distinct object of study and thereby 'rural studies' as a special area of study, (ii) familial and cultural attachments to the countryside, and (iii) the differential socialisation of new rural researchers and critical social theorists. He also suggest that critical approaches in rural studies are in their 'infancy' and there is a great danger that 'the theoretical strides issuing forth from critical social theory' will be either ignored completely or incorporated in such a manner that any uncomfortable and challenging elements can be ignored and the rural researcher can 'carry on much as before' (Cloke, 1989, p. 170). This possibility is not restricted to rural studies; Gregory, for example, has warned that

> there is a danger that attempts to assimilate critical insights into the corpus of traditional geography will leave its foundations undisturbed and its primary allegiances unchallenged: negation can follow hard on the heels of recognition, no matter how elaborate the ceremony, and the possibility ought not to be taken lightly.
>
> (Gregory, 1978, p. 157)

This paper will seek to explain why the notion of 'critical theory' should not be taken lightly, and also suggest that the current emphasis of critical rural studies on some form of a 'political economic' approach might usefully be supplemented with a recognition of the 'communicative' aspects of rural life. To achieve both these tasks the paper will focus on some of the arguments which have been put forward by the German philosopher/social theorist Jürgen Habermas. In the section that follows the nature of his ideas about critical social theory will be summarised with the aim of highlighting some of the basic but important issues raised within Habermas's work. After this attention will move to considering how some of his more specific arguments and concepts can be usefully employed within rural studies. Following this, attention will return once more to the general arguments of Habermas concerning the nature of critical social theory. In particular the paper will explore the issues surrounding what makes an interpretation of the countryside 'critical'. Before this can be done, however, it is necessary to outline what Habermas means by 'critical social theory'.

HABERMAS AND CRITICAL SOCIAL THEORY

> A theory of society can perhaps provide a perspective, can offer – to put it cautiously – hopes and starting points for the conquest of unhappiness and misery which are generated by the structure of social life. But it can do nothing to overcome the fundamental perils of human existence – such as guilt, loneliness, sickness and death. You could say that social theory offers no consolation, has no bearing on the individual's need for salvation. Marxist hopes are . . . directed towards a collective project, and hold out to the individual only the vague prospect that forms of life with greater solidarity will be able to eradicate, or at least diminish, that element of guilt, loneliness, fear of sickness and death.
>
> (Habermas in Dews, 1986, pp. 53–4)

This passage from a response to a question on the role of social theory can be seen to encapsulate many of Habermas's ideas on the aims, prospects and problems of critical social theory. As the passage from Habermas quoted above indicates, 'critical theory' can be said to be a theory or framework of understanding which offers and seeks to remove unnecessary social constraint: that is constraints which stem from the structures of social life. This idea of critical theory can be seen to have its roots in Marxism and the notion of establishing social theory with 'practical intent': where 'practical' is seen 'not in the sense of possessing a technological potential but in the sense of being orientated to enlightenment and emancipation' (McCarthy, 1987, p. 126). In a sense Robinson's charge that critical theory is 'utopian' accords with the notion of 'theory with practical intent' in that it is a theory which takes seriously the possibility of a better, more emancipated, future. On the other hand, if the use of the term utopian is seen to imply naïve or abstract idealism, then its application to the term critical theory should be rejected. Habermas and other well-known exponents of critical theory such as Adorno, Horkheimer and Marcuse have all been aware of the forces ranged against creating such a better life in practice.

It is very easy to buckle under in the face of such forces, and indeed the history of critical theory has many examples of people who are seen by some to have succumbed to these pressures. Perhaps the most notable was Adorno who, during the course of what he himself termed his 'shattered life' (Adorno, 1974), moved from being an advocate of clearly 'practical theory' to the notion 'negative dialectics' and thereby, so Habermas argues, 'gave himself over to the negativism of a thinking that . . . in the solitary experience of a self-denying philosophy, going round in its own aporias, [could see] the only possibility . . . of reason – however powerless – disguised in esoteric art' (Habermas, 1982, p. 232). Much of the labour of critical theorists since Adorno has been put into explaining why, even in the face of heavy pressures and possible failures, theory has to remain practical, has to remain committed to the achievement of the fullest possible degree of emancipation.

The way in which this 'justification' or 'grounding' of an emancipatory desire has been undertaken has varied considerably, even within the writings of individual critical theorists.

So-called 'linguistic turns' in social theory have become commonplace and Habermas's emphasis on communicative practice is frequently portrayed as an example of this general movement towards suggesting that society can be understood as a language (e.g. see Roderick, 1986; Taylor, 1991). However, Habermas has focused his attention on language for very specific and unique reasons. Habermas does not see language as being the only, nor indeed necessarily the major constituent of social life. His interest in language centres on its qualitative characteristics rather than its quantitative significance to social life. Habermas discerns in language qualities which, at least for him, provide a universal grounding for a concern for emancipation.

This argument, expounded in Habermas's 'universal pragmatics' (see Thompson (1984) for a good discussion of this part of Habermas's work), can be said, albeit simplifying greatly, to have three stages. First, Habermas argues that in our everyday communicative practice we necessarily, although often implicitly, raise and use 'truth' or 'validity' claims about ourselves and the world in which we live. Habermas then goes on to consider how people come to accept some validity claims and reject others. His answer to this issue is that the use of language in everyday conversation presupposes, and sometimes enacts, an 'argumentative procedure' in which reasons for accepting or rejecting validity claims are advanced, debated and evaluated. This argumentative procedure is 'the voice of reason, that we cannot avoid using . . . in everyday communicative practice' (Habermas, 1991, p. 244) mentioned above. The final step in Habermas's argument is to suggest that this voice of reason both presupposes and holds out hope for emancipation.

These arguments contained within Habermas's universal pragmatics are highly complex: each of the three steps of the argument requiring careful exposition (see Habermas, 1979, 1984, 1987b, 1990, 1992; Phillips, 1991b, forthcoming b; Thompson, 1984; White, 1988). In addition Habermas has been keen to explore the implications of his analysis of language for philosophy and ethics (see Habermas, 1987a, 1990, 1992, 1993) and for social analysis (see Habermas, 1979, 1984, 1987b). Clearly not all of the issues of relevance to, and following on from, Habermas's universal pragmatics can be discussed here. Instead this paper will concentrate on some issues which are of particular consequence to the study of rural change. Specifically the next part of the paper will use some of the ideas developed by Habermas to present an interpretation of social life in countryside change. This interpretation will draw upon some explicitly stated 'truth claims' – or concepts – largely drawn from the work of Habermas. The following part of the paper will suggest that the making of validity claims and the exercise of judgement are inevitable aspects of writing an account of the countryside (and indeed of writing an account of anything). The key issue for those writing on the countryside is therefore not how to avoid making claims to validity but what claims to make and on what grounds are people accepting or rejecting the truth claims of others.

HABERMAS AND THE STUDY OF RURAL SOCIAL CHANGE

Ideas from the early work of Habermas

One of Habermas's earliest writings was his *Strukturwandel der Öffentlichkeit* which was originally written as a post-doctoral thesis and was published in German in 1962 but only translated into English in 1989 under the title *The Structural Transformation of the Public Sphere*. This book provides a useful starting point for discussing Habermas's work in that it was not only his first major piece of work but also touches upon many of the issues on which he has subsequently worked. Indeed, Thompson (1990, p. 110) has argued that *The Structural Transformation of the Public Sphere* is 'crucial for understanding Habermas's most recent writings', although he also comments that this book did not receive great attention in the English-speaking world for many years.

The lack of attention directed to Habermas's early work on what he termed the *Öffenlichkeit*, which has been translated variously as 'public sphere' or 'bourgeois public sphere', is perhaps surprising in that notions of 'private' and 'public' on which it concentrates have become central to debates running across the historical, legal, literary, media, political, social and psychological sciences (McCarthy, 1989, p. xiii), and in the case of geography have been highlighted by feminist geographers (e.g. McDowell, 1983; MacKenzie, 1989; Rose, 1991, 1993). It has been suggested by several commentators that public/private divisions have been amongst the most fundamental of all social divisions (e.g. Delphy and Leonard, 1986; Pateman, 1983; Siltanen and Stanworth, 1984; Stacey, 1981) and Habermas certainly recognised that public–private distinctions could be seen as slicing their way across most of 'Western history', at least from Classical Greece (see Habermas, 1989b, pp. 3–4). Habermas, however, argued that the notions of public and private became particularly important in seventeenth and eighteenth century Europe with the rise to dominance of capitalist economies and constitutional states. He claimed that a public sphere, constituted as 'a sphere of private people coming together as a public' (Habermas, 1989b, p. 27), emerged in relation to, and sometimes in confrontation with, a privatised market and increasingly capitalist economy and a public realm of authority and administration.

In a sense Habermas's notion of the 'public sphere', as used here, may be taken to be an equivalent to the concept of 'civil society' as espoused by people such as Urry (1981): a sphere outside the state and outside the capitalist economy. However, the term public sphere may well be considered preferable for at least three reasons. First, as Goodwin (1989) has argued, the concept of civil society is rather a negative one: it becomes defined as anything that is not in the economy or in the state. Habermas's notion of a public sphere is much more positive in that it is a concept with a specific substance: the 'argumentative procedure' for accepting or rejecting validity claims. As Thompson puts it,

> if some aspects of the bourgeois public sphere were a veiled and disingenuous expression of class interests, nevertheless it embodied, argues Habermas, ideas and principals that went beyond the restricted historical forms in which it was realised. . . . It embodied what Habermas describes as a principle of 'publicness' or 'publicity' . . . namely that the personal opinions of private individuals could evolve into a public opinion through rational-critical debate of a public of citizens which was open to all and free from domination.
>
> (Thompson, 1990, p. 122)

For Habermas, the bourgeois public sphere was when it was established, and still is today, an ideology, by which he means an idea which is at the same time 'a socially necessary . . . falsity' and 'a claim to truth in as much as it transcends the status quo in utopian fashion' (Habermas, 1989b, p. 88). His argument for the public sphere as a 'socially necessary falsity' was that the principle of organising social life on the basis of open and unconstrained discussion was established by the eighteenth century bourgeoisie because they were at one and the same time excluded from the dominant political institutions and caught up in the rapid social changes associated with emergent capitalist social relations. The ideal of a public sphere held particular appeal for this group: it was by definition easy for them to enter (particularly if the notion of 'open to all' was phrased to mean open to all property-owning men), and once linked to state power through 'rational policy' it could be used to organise social life for their particular interests (see Corrigan and Sayer (1985) for similar arguments). In this way the concept of the public sphere was, as Habermas (1989b) puts it, 'a category of bourgeois society'. However, whilst outlining bourgeois origins for the concept of the public sphere, Habermas is at the same time claiming that it is an idea which has more universal value: namely that it establishes the principle that emancipation of people implicated within discursively established or supported relations of domination can follow only from open and unconstrained debate.

A second benefit of Habermas's conception of the public sphere over Urry's notion of civil society is that Habermas uses his concept in a very relational way. For instance, Habermas sees the formation and operation of the public sphere as being related to a market economy and the bureaucracy of state power, as well as being related, albeit frequently only immanently, to actions orientated to reaching mutual understanding such as democratic decision-making and critical debate. Habermas also highlights how relations between the market economy, the state and the public sphere could change. Specifically he argues that transformations of state power, the rise of large industrial complexes and the emergence of media for mass communication during the course of the nineteenth and twentieth centuries led to fragmentation of the public sphere into publicity, public relations work and public opinion research.

A third potential attraction of Habermas's conception of the public sphere, at least for geographers, is that it can be readily spatialised. Although it is correct to suggest that Habermas's arguments imply that there can be 'as many publics as there are controversial general debates about the validity of norms' and that public spheres 'come into existence whenever and wherever' people enter into debates evaluating the validity of 'social and political norms of action' (Benhabib, 1992, p. 105), it is also the case that Habermas linked the public sphere to highly territorialised types of spaces, including the household, the 'salon', the 'town' and the 'court'. This aspect of Habermas's work suggests that recent criticisms concerning a lack of any geographical imagination (e.g. Gregory, 1989, 1991; Howell, 1993) are overdrawn, although it is certainly the case that the spatial dimensions of the concept of public sphere can be outlined much more explicitly than was done by Habermas in his *Structural Transformation of the Public Sphere*. Howell (1993, p. 314), for example, has suggested that the work of Hannah Arendt on the public sphere has a 'spatial quality' about it, while Benhabib (1992) has provided a highly interesting discussion of public spaces which draws on the work of Habermas, Arendt and liberal theorists such as Ackerman (see Figure 2.1). Although her purpose was to examine differences in public space from the point of ethical and political theory, Benhabib does recognise that concepts of public space have many other connections. Human geographers in particular could explore some of these connections – and indeed a small start to this will be undertaken in the next section of this chapter (see also Ogborn, 1993; Phillips, forthcoming a) – although one might add that a critical human geography would need

Conceptions of public spheres	**Legalistic**	**Agnostic**	**Discursive**
Proponent of concept	**Ackerman**	**Arendt**	**Habermas**
Forms of public space	**Space of socially neutral conversation** (Only non-contestatory issues are allowed to be subject of public debate) **Rule bound space** **Orderly space**	**Space of appearance** ('moral and political greatness, heroism and pre-eminence are revealed, displayed and shared with others') **Competitive space** (People compete for recognition, precedence and acclaim) **Space of nuturing of selves** (Meanings circulated which 'guarantee protection from, futility of individual life')	**Space of democracy** (Place where people affected by social norms and collective political decisions have a say in their formulation) **Space of debate about contested validity claims**

Figure 2.1 Conceptions of public spaces. Source: Based on Benhabib, S. (1992) *Situating the Self: Gender, Community and Postmodernism in Contemporary Ethics*, Polity, Cambridge.

to be aware of the importance of retaining both the 'utopian' and 'relational' aspects of Habermas's notion of the public sphere and not simply emphasise the issue of spatiality.

This is not to say that Habermas's notion of the public sphere is above criticism. Thompson (1990) has, for example, argued that Habermas tends to focus his attention on rather superficial aspects of culture and politics within contemporary societies and assumes rather than demonstrates that people are passive recipients of media messages. Thompson also highlights some terminological problems associated with the notion of a public sphere. In particular he notes two distinct ways in which the private–public dichotomy has been commonly used: first, to differentiate out a domain of institutionalised political power associated with a sovereign state from the so-called private concerns of a market economy and personal or familial relations, and second, to differen-

Private realm	Public sphere	Sphere of public authority
Civil society (realm of commodity exchange and social labour)	Public sphere in the political realm	State (realm of the 'police')
	Public sphere in the world of letters (clubs, press)	
Conjugal family's internal space (bourgeois intellectuals)	(Markets of culture products) 'Town'	Court (courtly-noble society)

Figure 2.2 Habermas's representation of the bourgeois public sphere in eighteenth century Europe. Source: Habermas, J. (1989) *The Structural Transformation of the Public Sphere: An Inquiry into a Category of Bourgeois Society*, Polity, Cambridge.

tiate a sphere of 'publicness' from 'privacy'. In this second sense, the public sphere means 'what is open for all (or many) to hear or hear about' (Thompson, 1990, p. 240) and the private is 'what is hidden from view'. Habermas recognises both these distinctions (see Figure 2.2) but tends to slide them into a series of more 'concrete' or 'empirical' categories, such as the conjugal family, the salon, the town, the police and the court. This running of theory into the empirical is one of the strengths of Habermas's *Structural Transformation of the Public Sphere*, but it is also a potential cause of misunderstanding in that the different senses of public and private are at times obscured. One way of preventing this particular problem might be to employ a distinction of official, intermediate and unofficial spheres (see Fraser, 1989, p. 118; Corrigan and Sayer, 1985, p. 12; and Stacey, 1981, for similar divisions) as well as public and private spheres. The first set of distinctions will be used to indicate the extent to which a particular practice and associated social relations operate through institutionalised power, while the second division focuses on the extent to which practices are visible and accessible to others. Figure 2.3 illustrates how one might combine the two sets of divisions.

The purpose of this paper is not just to review the work of Habermas but also to explore how it may be of value to understanding and developing a critical study of rural change. The next section will explore how the notions of public and private spheres, together with those of official, intermediate and unofficial spheres, might provide fruitful avenues for future rural research.

Public and private spheres as a perspective on existing rural studies?

Habermas (1989b) refers to the public sphere as both an historical object of analysis and as a 'category of bourgeois thought'. Likewise, it will be claimed

	Public sphere	Private sphere
Official sphere	State (administrative capital) Mass media (cultural capital)	Market economy (economic capital)
Intermediate sphere	Interpersonal communication	Social economy (social labour)
Unofficial sphere	Nurturing of 'selves'	Informal economy (reproduction of bodies/labour power)

Figure 2.3 A representation of official/unofficial and public/private spheres for the late twentieth century.

here that it is possible to use the ideas of public/private spheres and official/unofficial/intermediate spheres both as constituents of rural societies worthy of analysis and also as a category of thought employed in rural studies. Indeed, this section will demonstrate how studies on rural change have continually 'bumped up against' private/public and official/unofficial distinctions but have largely failed to recognise these divisions as important constituents of analysis, and that rural studies have tended to fall into a category of thought largely defined by these divisions.

A useful starting point in demonstrating these two arguments is the nature of existing critical rural research. As already mentioned, most of the rural studies which have been framed in terms of social critique have utilised a 'political economy' perspective. While there has been some debate about the exact parameters of such an approach (see for example, Bradley, 1981; Cloke, 1989; Marsden, 1988; Marsden *et al.*, 1986), a common feature is an emphasis on 'property and labour within the context of generalised circulation of commodities' (Friedmann, 1986, see also Whatmore, 1991). Critical rural research in the main has focused on the official sphere, that is on the market economy and the administrative state, although there has been some recent recognition of the need to recognise some more unofficial dimensions of social life, such as 'unproductive' labour and 'patriarchal gender relations' (Whatmore, 1991; see also Cloke and Phillips, 1991; Phillips, 1991a).

Some of the disagreements about the best form for a 'political economy of the rural' can be seen to reflect divisions between public and private and between official and intermediate spheres. As O'Connor has argued, for example, liberal forms of political economy – such as neoclassical economics and Weberian class analysis – seek to explain social actions largely through market relations, which are, he suggests, the 'most objectified and formal level of capitalist society' (O'Connor, 1987, p. 5). Marxist political economy, in his view, is 'more "totalistic" than market theory' in that it moves beyond the formalised categories of bourgeois society and into the realm of 'social labor [*sic*] and social-class relations' (O'Connor, 1987, p. 5). In other words, while liberal political economy focuses on officially sanctioned relations between individuals operating as autonomous market agents, Marxist political economy recognises that people are also connected by less official relations, including those of labour discipline and productive cooperation. Marxist political economy can be seen to recognise both official/private and unofficial/private relations. However, for O'Connor, even this expanded gaze is insufficient:

> both 'market relations' and 'social labor' [*sic*] are . . . abstractions at both the social and philosophical levels. Nowhere in the world are there pure exchange relations or processes of social labor . . . which are not inscribed and structured by cultural, ideological, and other 'social productive forces'. Culture and ideology are embedded in market and production forces and relations in complex ways. . . . History and class struggle, therefore, are not structured by movements of social labor and capital alone, nor less by changes in wages, prices, and profits, or 'market forces'. They are ambiguously structured by culture and ideology, tradition and fantasy, personality development, and other social process which cannot be reduced to material life strictly defined.
>
> (O'Connor, 1987, p. 8)

A critical theory for O'Connor has to expand its focus even more than Marxist political economy does, and should include issues such as personal identity and fantasy which may be considered to be highly private – both in the sense of being highly individual and also by being hidden from view. For O'Connor at least, critical theory needs to encompass the private/unofficial worlds of self-identity.

This argument poses a challenge to those who see either a liberal or a Marxist political economy as a sufficient basis from which to understand rural social change. Indeed, given that much 'critical' rural studies has remained fixated on the delineation of landed property rights (e.g. Flynn, Lowe and Cox, 1990; Lowe, Marsden and Munton, 1990; Marsden and Murdoch, 1990; Whatmore, Munton and Marsden, 1992) and on the impacts – rather than the social constitution – of government policies, one can suggest that critical rural studies has been constructed almost exclusively within the 'official private sphere'.

There are exceptions to this: Cloke and Little (1990) have begun to explore the social construction of rural planning and policy making, while Redclift (1985), Little (1987), Pile (1990b, 1991b) and particularly Whatmore (1991) have argued for the need to move beyond the realm of commodity production. Particular attention is paid in these writings to the way concepts such as work have been viewed in very narrow, and highly gendered ways, and how 'work' in the home and other often unrecognised locations may have important implications for understanding actions, including those which are seen to be conducted in the official realm of the workplace. However, while such studies can be said to mark an important advance into the intermediate/private realm of work, one can suggest that this analysis still remains bounded by certain public/private and official/unofficial divisions. This point is hinted at by Whatmore in the conclusion to her study *Farming Women*:

> [there are] two important limitations . . . [in] the kind of analysis that I have undertaken. One is that, in focusing on the regime of the family household, the attention paid to the wider gender order as it influences that regime has been restricted. . . . A second is that the material relations of sexuality, and emotional and physical violence, which are constitutive of patriarchal relations, require more direct and sustained attention than they have received here.
>
> (Whatmore, 1991, p. 142)

In other words, neither the role of gender images in the official public sphere nor the significance of the unofficial and/or private relations between household members were incorporated in the study as fully as they might have been. Whatmore's study is not alone in these omissions: indeed its distinctiveness lies in its recognition of the potential significance of these issues.

The neglect of the unofficial/private sphere and the very partial attention paid to the official public sphere – where attention is largely directed towards the economic and political as opposed to the social and cultural – can be accounted for in a variety of ways. One important factor may well be the social character of those undertaking rural studies: given that most rural researchers are middle-class males it is arguably no coincidence that emphasis has been placed on labour which brings remunerative rewards and that issues such as sexual violence have been ignored. One can also suggest that the partial nature of critical rural studies is related to the constitution of public and private spheres.

For instance, within rural studies there has been a deep division between the critical political economy perspective and what has been termed 'community studies' (see Bell and Newby, 1973; Harper, 1989; Wright, 1992, for reviews of this approach). It is possible to suggest that this division has been so persistent, and indeed has been so hard to transgress, because it embodies divisions between and within public and private spheres. While the gaze of critical rural studies has largely concentrated on the official and intermediate private spheres and selected parts of the official public sphere, advocates of community studies can be seen to have focused on the intermediate public sphere to the exclusion of any consideration of the private sphere. This point has been well made by Wright who has argued that community studies, such as those of Rees (1950) and Littlejohn (1963), focus on a 'public space' or 'arena of social activity' which lies between, 'on the one hand, the household, and on the other, the state' (Wright, 1992, p. 202). Wright clearly identifies both the character and the limitations of this perspective in these comments on the work of Rees:

> On households he gives plenty of ground plans of houses, but leaves tantalising gaps in his account of how space is used during the family life cycle. He describes the farm work but never explains how it is organised. There are two paragraphs on women in the household. They say there is

> a division of labour between the sexes and the activities of women are largely to the house and farm yard. In addition to her household duties, she raises poultry and makes butter. She markets the products herself, and on the proceeds she runs the house and clothes herself and the children. . . . Rees seems not to have asked equivalent questions of the men, about their side of farm production and what they spend the proceeds on. . . . On relations between the community and the state, Rees' monograph keeps the political context of anglicisation and state policy in view, but this serves to define the public space he is interested in. The community life of these scattered farms takes place when visitors gather by the hearth of a family's living room. Then what had been private space is transformed into a public community area. He considers this community, which is situationally created out of the living room in each house, to be threatened by encroaching English and urban standards which equate community with a nucleated settlement crowned by a public hall.
>
> (Wright, 1992, pp. 204–5)

Rees directs attention at a public sphere lying between household and state: he effectively therefore explores what here is called an intermediate public sphere. This is an arena which although not driven directly by forces of state power and the market economy is clearly conditioned by them, and which although linked into the personalities and identities of individual people is also not reducible down to them. This intermediate public sphere may be of crucial significance, but Rees's analysis effectively serves to represent this sphere in isolation from other spheres. For example, Rees does not delve into the social divisions and relations which operate within households, nor does he outline the constitution of forces driving anglicisation, both in terms of the administrative and cultural capital involved in state policies nor the links to a capitalist market economy. Indeed, Rees's study never really penetrates beyond the most official level of the private sphere: the market relations of agriculture are dealt with but not the social organisation of work. Furthermore, Rees establishes clear gender and racial divisions through his construction of a seemingly autonomous 'intermediate sphere'. As Wright comments, there are only two paragraph's on women in Rees's account: he clearly saw them as being marginal to the 'community' he was studying. Indeed women were placed firmly into the private/unofficial world of domesticity or at the margins of the private/official world of commerce. Similarly, the English are placed firmly in the official spheres of money and government policy.

This brief discussion of political economy and community studies has suggested that while one can see public/private and official/unofficial divisions within both perspectives, each perspective has focused on one or two divisions (see Figure 2.4). This has effectively acted to reproduce and naturalise these divisions and has led to the divisions themselves not being open to scrutiny. In the next section attention will be directed towards showing how the links between public/private and official/unofficial social spheres might be considered and how hitherto ignored aspects of these divisions could usefully become new objects of concern within critical rural studies.

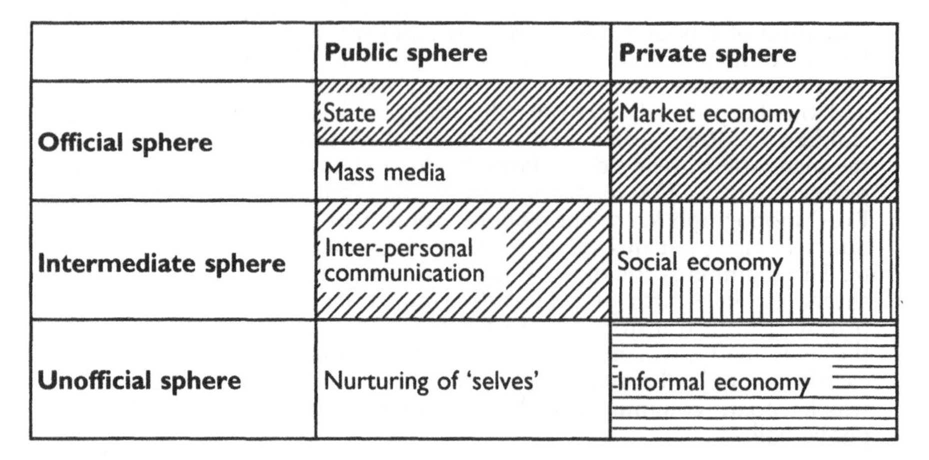

Figure 2.4 Rural studies and official/unofficial and public/private divisions. ▨ Political economy (Marxist and Weberian); ◫ political economy – Marxist; ⊟ political economy – feminist; ▨ community studies.

Beyond public/private dichotomies: examples from past rural research

In order to illustrate how public/private distinctions are significant in understanding rural social change it is useful to return to some of the earliest examples of critical rural studies and in particular to the work of Howard Newby. While many of most well-known advocates of a political economic approach to rural studies (e.g. Lowe, Marsden and Munton, 1990; Marsden, 1992; Marsden *et al.*, 1986; Whatmore, Munton and Marsden, 1992) have focused on the arguments of Newby concerning the centrality of property relations, some of the other arguments developed by Newby point beyond this private/official sphere and illuminate significance of divisions within and between the public and private spheres.

For example, in his book *The Deferential Worker*, Newby suggests that nineteenth century villages were composed of two distinct 'communities': the 'dark village' and the 'official village' (Newby, 1977, p. 46; see also Hobsbawn and Rudé, 1969). The dark village was the cultural world of the agricultural working class and was marked, so Newby argued,

> by a fatalistic acceptance; an accommodation to a situation which the agricultural worker was powerless to alter and to which he [*sic*] reacted to the most part by . . . withdrawal into a close knit . . . circle of fellow workers. From this he would emerge to behave to his 'betters' with the appropriately servile and 'deferential' demeanour that was expected of him.
>
> (Newby, 1977, p. 28)

The official village was the cultural world of country landowners, farmers, country clergy and multifarious other middle-class rural residents.

These so-called 'dark' and 'official' villages can be seen to have formed distinct yet mutually constitutive private and public spheres. The dark village, for example, was a private sphere in the sense that it was both unofficial and private. It was unofficial in that its precepts were defined without recourse to, and arguably in defiance of, the powers of the state and a market economy. It was private in the sense that it was hidden from view: it was a 'habitus' which the farm worker had to leave behind in order to converse with the landowner, the farmer, the priest or any other member of the official village. This official village was a public sphere in that it was supported by the institutions of market and state and because it involved show (Newby, 1977; Howkins, 1991), or using Habermas's words, it was a sphere of public representation in which, 'the "ranks" paraded themselves, and the people applauded' (Habermas, 1989b, p. 38). One might add that at least some elements of this sphere, such as agricultural societies (see Wilmot, 1990) and the reading salons, might be considered to constitute a 'public sphere' in the sense that they involved reason: that is they involved, at least in principle, debates to resolve contested validity claims. Needless to say they also involved a variety of more private concerns as well.

It is important to note that these public and private spheres were not autonomously created, but each served to reproduce the other through systems of 'deference' and 'paternalism' (see Newby, 1977). Furthermore, as Howkins (1991) has clearly demonstrated, the distinctions between these private and public spheres were not static: they had to be reproduced and were subject to several re-workings and transformations. An important task for a critical rural geography might be to outline the precise historical geographies of the changing public/private divisions. Indeed, whilst several historians and historical sociologists have begun to examine, revise and reformulate Habermas's historical account of the bourgeois public sphere (see Baker, 1992; Eley, 1992; Landes, 1988; Ryan, 1992; Schudson, 1992; Zaret, 1992), little attention has been paid to his geographical account, particularly at scales below the nation-state. *The Structural Transformation of the Public Sphere* is a very urban centred account: Habermas talks, for example, of 'the court' and of 'the town'. Many of his critics also produce urban centred accounts: Landes, Eley and Ryan, for example, argue variously for the inclusion of 'the crowd', 'the street' and 'the square' in discussions of the public sphere. The rise of the public sphere is hence often written as a growth in the urban, but does it have to be written in this way? There clearly are public spaces within the countryside – the village green, the village fair, the parish hall, the country pub, the meeting in the fields – and also a range of private spheres. Do these public and private spaces have direct urban counterparts or are they given particular rural inflexions? If, as Warner (1992) argues, participation in a public sphere requires a suspension of one's private identity, then one might speculate that there might be fewer unremittingly 'public' public spheres in rural communities because it may be much harder to become an anonymous subject: a person with no private history. Conversely, one might suggest that it is much harder for people in the countryside to retreat into a completely private sphere in that the village publics regularly call on people to become active in a community or may persistently act as service providers in the form of 'good' neighbours. On the other hand, rural/urban differences in public/private spheres might be of minimal significance: rising privatism and familism – home and the family-centred lifestyles – has, for example, been identified in both rural (Newby *et al.*, 1985; Cloke and Thrift, 1990) and non-rural contexts (Gorz, 1982; Marshall *et al.*, 1988). A geography of rural public and private spheres might provide a very valuable window on Habermas's *Structural Transformation of the Public Sphere*. In the present context, however, all that will be attempted is a discussion of some rural studies which can serve to highlight the significance of thinking about public/private divisions in periods beyond the nineteenth century.

Newby, along with Colin Bell and Peter Saunders, was involved in writing *Property, Paternalism and Power* which can be read as revealing the significance of public/private distinctions, and indeed may indicate how the public and private spheres of the official and dark villages of the nineteenth century have been transformed. Newby *et al.* (1978) argue that during the late twentieth century 'village communities' became transformed and either 'encapsulated' or 'farm-centred' communities established. In the former, remnants of the official and dark villages have become fused and centralised into a 'self-contained and tightly knit group' (Newby *et al.*, 1978, p. 194), often concentrated in one area of a village. In the latter, farmers and farm workers are seen to withdraw from the village and reside on, or close to, their farms. Both communities reflect, so Newby *et al.* argue, a change in the way farmers behave: farmers are, they suggest, increasingly concerned 'to maintain their control of workers on the farm and . . . to protect what they see as the wider political interests of farmers at the district or county levels' (Newby *et al.*, 1978, p. 195). One result is, so Newby *et al.* claim, that control of the local village, so significant in the period of the 'official village', is now seen as irrelevant.

If one accepts this interpretation of changes in village life then one could suggest that what Newby *et al.* are talking about is the withdrawal by farmers from the 'intermediate public sphere' of the village into the 'private' domain of the farm and the 'official public sphere' of local authority (and higher authority – see Cox and Lowe, 1984; Pile, 1990a, 1990b, 1991a) politics. *Property, Paternalism and Power* effectively follows the farmer into the private domain and into the official public sphere, concentrating as it does on farmers' control of the farm worker and on farmers' influence in local government. The book does, however, raise a further point of interest – namely what happens to the intermediate public sphere of the village when the farmers withdraw from it? One answer, provided by Newby *et al.*, is that the 'village community' becomes replaced by a 'communion of localism' (Newby *et al.*, 1978, p. 195). By this they mean a 'community of feeling' amongst and about being 'locals':

> the village retains an important *symbolic* significance. Recent changes in the social composition of the village allow, for example, both farmers and farm workers to feel misunderstood and to be threatened by outside forces . . . there was talk of 'invasion' of aliens and of 'natives' . . . new-

> comers . . . seemed to provide a convenient scapegoat for any recent undesirable change.
>
> (Newby *et al.*, 1978, pp. 195–204)

Newby *et al.* (1978) argue that during the second half of the twentieth century farmers and farm workers came not only to see themselves as a 'native' rural population under threat but also increasingly came to interact socially within a localised farming community – a social world of 'markets, agricultural societies and shows, cocktail parties and so on' (Newby *et al.*, 1978, p. 210; see also Carr and Tait, 1990, 1991, for similar arguments). In other words, one can suggest that farmers were replacing the intermediate public sphere of the official village with a narrower and more unofficial public sphere: that of the local community.

The rise of localism can be seen to have had a profound influence on rural societies: the 'local' and the 'incomer' have framed discussions at both the theoretical level and in the establishment of policy initiatives (see Cloke and Little, 1990). For some researchers, such as Harper (1987), the local–incomer category appears to be the central form of social 'difference' (Philo, 1992, p. 200). However, as Cloke and Little (1990, p. 11) argue, the local–newcomer division may well be better conceptualised as 'a specific localized symptom of more general causative social processes'. It may well be that the processes establishing the local–newcomer distinction lie within and across a range of public and private spheres. For example, Cloke and Little (1990), and indeed Newby *et al.* (1978), argue that the local–newcomer division is the outcome of economic restructuring and associated social recomposition. In other words the origin of the local–newcomer distinction lies within this interpretation in the official and intermediate private spheres of a capitalist economy. Many advocates of community studies, on the other hand, have placed particular, and arguably too much, emphasis on the 'intermediate public sphere' of social interactions. Furthermore, the newcomer–local division has reached the public sphere of the media: not least in the characters of the BBC radio programme *The Archers*. It is interesting to speculate about the extent to which the acceptance of local–incomer divisions in the 'official public sphere' has served to reproduce the division within villages. Cloke and Milbourne (1992) have certainly suggested that the national circulation of officially sanctioned images of the countryside may be an important constituent of local cultural conflict. The work of people such as Mahar (1991) has also highlighted the way unofficial and 'feminised' activities – of caring, of organising, of serving – may be crucial in establishing rural communities of distinction, while Middleton (1984) and Little (1987) have indicated how participation in public and private spheres are highly inter-related and heavily gendered. Indeed it is possible to discern a clear gendering of both private/public and official/unofficial divisions, a gendering which has its own localism:

> men routinely watch national news reports whereas women expressed little or no interest in these programmes. But they were deeply concerned with local news reports. National news, concerned essentially with 'public' economic and political events has little practical relevance for wom-

> en's daily lives but local news served a vital role in enabling women to monitor events in the locality. All the women in our groups expressed great anxiety about the dangers of sexual attack in public open spaces and were deeply concerned both for themselves and their children. ... Their fears are amplified by reports of rapes, murders and child abduction in both national newspapers and the local media, creating a situation in which it becomes necessary to mediate personal desires for contact ... with fears of potential danger. It is tempting to speculate about the marked tendency for women to become involved in local community politics ... rather than at the national level.
>
> (Burgess, 1990, p. 156)

In other words, many women appear to inhabit a local, private, unofficial world while men try to connect to a more national, public and official sphere.

A cautionary remark may be necessary: it may appear that it is being suggested that the public/private and official/unofficial distinctions solve all the problems of rural studies. This is not the argument: rather, it is being suggested that there is a need to recognise processes operating in both public/private and official/unofficial spheres and to consider their interaction. In a sense the twin sets of distinctions serve essentially as a reminder of how complex and multi-dimensional societies can be: that one may need to look beyond the most immediate of distinctions to consider if there are unrecognised and/or more or less 'collective' influences at work. These reminders may seem rather inconsequential, but they can be seen to embody some of the central features of Habermas's notion of a public sphere and indeed his arguments about the distinctiveness of a critical social theory.

What is so critical about a public sphere?

The above section has argued that one failing of rural studies has been a failure to think across public and private divisions. It was suggested that this is very evident in research on localism. A further criticism that can be levelled at some rural research is that it has rather tended – through an emphasis on notions of 'belonging', 'boundedness' and 'difference' (see Cohen, 1982, 1986; Harper, 1987) – to celebrate the exclusionary practices of localism: valuing 'belonging' and 'boundedness' can lead into indifference to outsiders/incomers, while as being seen as 'different' has for many people not been a source of pleasure but the start of oppression and domination. The notion of a communion of localism forming a public sphere in Habermas's strictest sense of the term is misleading. Indeed, as Murdoch and Marsden (1991, p. 50) argue, processes such as commodification and privatisation may well be making the countryside 'not yet private but not accessible to many'. This might suggest that Habermas's conception of public/private spheres has somewhat limited applicability to rural studies. There are, however, two points that need to be made to demonstrate that this is not the case. First, it is important to note that Habermas himself suggested that a public sphere open to all has not been realised (Habermas, 1989b), even during the eighteenth century when arguably a close approximation to a public sphere emerged (although see Baker,

1992; Eley, 1992; Landes, 1988; Ryan, 1992; Stallybrass and White, 1986; on some of the far from emancipatory features of the public sphere at this time). Nevertheless, for Habermas it is important to recognise that there have been some more and less good approximations to an emancipatory public sphere (Habermas, 1989a, 1989b). Likewise it may be possible to distinguish public spheres in the rural realm which approximate, in some senses, to the notion of a public sphere as an arena in which people enact open and unconstrained discursive will-formation. Second, and in a critical sense rather more importantly, for Habermas the concept of a public sphere was not simply an empirically validated one but was also a 'practical' one in the sense that he argued that the *principles* of the public sphere underpinned many of the central, and for Habermas at least most, valuable institutions of modern Western societies: most directly democracy (Habermas, 1979, 1989b; McCarthy, 1991), but also academia (Habermas, 1974, 1989a) and a variety of social struggles and protests (Habermas, 1989a). Indeed, as McCarthy (1991, p. 193) notes, Habermas effectively claims that without the principles of the public sphere most of what might be considered to be 'rational practices' – practices which involve taking decisions after evaluating alternatives – 'would lose their sense', which in turn 'would entail far-reaching changes' in the form of life for much of contemporary society. Furthermore, for Habermas, such changes would not be for the better. Rather life completely devoid of reasoned argument would 'In Habermas's view . . . mean the end of our main alternative to violence, coercion, and manipulation as a means of conflict resolution and social coordination' (McCarthy, 1991, p. 195).

Violence, coercion and manipulation are already present within rural societies and many people would wish that they might be replaced by reasoned argument. For Habermas, however, the favouring of reasoned argument is not merely a matter of personal preference but is presupposed in the way people use language in everyday speech. This argument, developed as his theory of 'universal pragmatics', represents one of the key differences between Habermas and a wide range of other contemporary theorists. In the realm of political and ethical philosophy, for example, people such as Rawls have argued for the desirability of many of the principles of the public sphere, yet they seek to justify these principles only in terms of being central elements of current liberal political culture (see Rawls, 1980, and also Rorty, 1982, McCarthy, 1991). Of more direct consequence, Habermas's emphasis on language as a universal basis for reasoned argument has important implications for understanding critical academic activity.

WHAT ROLE FOR THE CRITICAL ACADEMIC OF THE RURAL?

Raising a few awkward questions for researchers of the rural

Mormont (1990, p. 36) has recently criticised rural researchers for failing to examine their own role in production of 'knowledge of the rural'. While he focuses specifically on differences in the way the rural has been conceptualised, there is also the issue of what constitutes knowledge. Rural researchers have been notably reluctant to address such issues, although there has been some evidence that at least some of its practitioners are concerned about some implications of a wider academic debate circulating around the term 'postmodernism' (see Murdoch and Pratt, 1993; Philo, 1993). One issue that this debate has raised is what precisely is the status of various forms of knowledge. In particular does the recognition of a diversity of ways of seeing the world mean that academic debate becomes 'just opinion swapping' (David Livingstone, quoted in Cloke, Philo and Sadler, 1991, p. 204). Most geographers seem to want to avoid such relativism (although see Doel (1992, 1993) and Chapter 3 of this volume), but the grounds for this resistance are rarely spelt out (see Strohmayer, 1993). One of the most distinctive features of Habermas's work is that he wants to suggest that it is both possible and desirable to see knowledge as more than mere story telling or opinion swapping. In particular Habermas suggests that knowledge has 'problem-solving capacities' (Habermas, 1987a): that is it enables people to do things.

The exact role that knowledge plays has been a continuing problematic for Habermas (see Phillips, forthcoming b), and indeed for earlier 'Critical Theorists' such as Horkheimer, Adorno and Marcuse (see Wohin, 1989). A key argument of Critical Theory, ever since the publication of Horkheimer's 'Traditional and critical theory' (Horkheimer, 1937), has been that intellectual processes are inextricably linked into social processes. This point has recently been highlighted in the geographical context by Unwin who comments: 'Many undergraduates, who have been socialised to see education as the acquisition of skills and knowledge that are seen as being useful by society, enter higher education in order to be better placed later to receive the rewards of such knowledge' (Unwin, 1992, p. 18). Knowledge becomes a credential: an institutionally guaranteed cultural competence which can be exchanged for economic and material reward (see Bourdieu, 1984, 1988; Cloke and Phillips, 1991; Thompson, 1990, for useful discussions of the concept of credentials).

Credentialism and the 'economic valorization' (Thompson, 1990, p. 155) of knowledge has implications for what constitutes knowledge. Unwin, for example, suggests that for people who see knowledge as simply a credential to realise economic rewards, the idea of questioning the status of knowledge is 'at the best meaningless, and at the worst positively damaging to their future careers' (Unwin, 1992, p. 18). In other words philosophising about the status of knowledge might undermine the credential which some people are striving to gain. His arguments have clear echoes of some rather earlier remarks of Habermas:

> In every conceivable case, the enterprise of knowledge at the university level influences the action-orientating self-understanding of students and the public. It cannot define itself with regard to society exclusively . . . to systems of purposive rationality. . . . The translation of scientific material into the educational process requires the very form of reflection that was

> once associated with philosophical consciousness.... This dimension must not be closed off.
>
> (Habermas, 1971, pp. 4–9)

The reason why knowledge should not be seen merely as a credential to be accumulated for economic exchange is, for Habermas, that knowledge is a means of empowerment and emancipation: 'in the end insight can coincide with emancipation from unrecognised dependences – that is, knowledge coincides with the fulfilment of the interest in liberation through knowledge' (Habermas, 1974, p. 9). In other words, the more knowledge a person has about their condition the more power they have at their disposal and the freer they are to make their own destinies. Sayer makes the point clearly:

> the point of all science, indeed all learning and reflection, is to change and develop our understandings and reduce illusion.... Learning, as the reduction of illusion and ignorance, can help free us from domination by hitherto unacknowledged constraints, dogmas and falsehoods ... what is learning for, if it is not to change people's understanding of the world and themselves.
>
> (Sayer, 1992, p. 253)

The production and acquisition of knowledge for Habermas and Sayer hence has an inherently critical function: it can change people's lives – both individually and collectively. The truth of knowledge for critical theorists lies not as a mirror or representation of things as they are, but rather in the extent to which it makes the world more intelligible to people and enables them to act in less constrained ways.

The claim that knowledge holds the potential to overcome relations of power has placed Habermas within 'the Enlightenment project' (see Cloke, Philo and Sadler, 1991, p. 188). The Enlightenment ideal that people and societies can and should be changed by the power of rational knowledge has, however, come under attack, most visibly, but one should add not exclusively, through notions of postmodernism (Cloke, Philo and Sadler, 1991, p. 188; Gregory, 1989; Habermas, 1987a; McCarthy, 1991). One line of criticism has been that of Foucault who argues that knowledge does not dissolve relations of power but is in itself a form of power (see Matless, 1992, and Chapter 1 of this volume). Rather than knowledge versus power as in Habermas, Foucault sees 'power/knowledge' (Foucault, 1980). Rational knowledge in this view does not simply shed 'enlightenment', but rather, 'subjects the world to a particular kind of illumination; one that casts shadows into which one cannot see, and so requires mechanisms of surveillance and control' (Pile and Rose, 1992, p. 126).

These criticisms of the Enlightenment project have great validity. The notion that acquisition of knowledge can overcome relations of power, for instance, has all too frequently been translated into logocentricism – the notion of individual and autonomous thinking subjects – and hence into ethnocentrism, sexism, technocentrism and credentialism (for useful recent geographical discussions of these 'isms' see Gregory, 1991; McDowell, 1992a; Pile and

Rose, 1992; Rogers, 1992; Slater, 1992). Bauman has suggested that the era of the Enlightenment was associated with state power to 'shape and administer the social system' (Bauman, 1987, p. 2), while Livingstone (1992, p. 126) has argued that this period saw geographers enveloping and even 'engulfing' increasing areas of the world within 'European scientific ways of seeing' and the practices and consequences of European colonialism.

What is less clear, however, is whether the ideas of Habermas and the concept of 'critical theory' is irrevocably linked to particularity in social vision and to practices of surveillance and control. There are at least two reasons for suggesting that this is not the case. First, Critical Theorists have long recognised the 'black side' (Habermas, 1987a, p. 106) of the Enlightenment project: indeed each of the major figures in the tradition have written on this very issue (i.e. Adorno and Horkheimer, 1979; Marcuse, 1964; Habermas, 1987a, 1989b). Second, Habermas has launched his own critique of postmodernists (see Habermas, 1987a, 1989a, 1990, 1992), arguing, amongst other things, that they see themselves as being in some sense above society, that they routinely undertake acts of 'performative contradiction' (Habermas, 1987a, p. xv) by putting forward 'truth claims' in the very act of uttering that there are no truths, and that they are effectively a socially conservative force in that 'they have abandoned any hopes of conscious social change' (Wohin, 1989, p. xxv). Against the arguments of postmodernists, Habermas seeks to present reasons for maintaining that: (i) all knowledge is socially (or more precisely 'intersubjectively') produced; (ii) even in everyday communicative practice, and certainly within the specialist academic discourses, people raise and use 'truth' or 'validity claims'; and (iii) knowledge can change things. These arguments of Habermas will now be outlined and their implications for the prospect of critical rural research examined.

The intersubjectivity of knowledge

Amongst the criticisms of the Enlightenment project which have been made by Critical Theorists for some time has been the notion of individual and autonomous thinking subjects. For Habermas, knowledge is 'intersubjectively' produced: that is it is the outcome of interactions between communicating individuals. Habermas has developed this argument in at least two distinct, although generally complementary, ways. First, in writings scattered throughout his career (e.g. chapters in Habermas, 1971, 1974, 1983, 1989a), he has linked the activities of the academic into his principles of a public sphere. In *Towards a Rational Society*, for example, he suggested that there was an 'affinity and inner relation' between 'the enterprise of knowledge on the university level to the democratic form of decision making' and that both research and teaching at universities should enact the principle of 'decisions . . . made dependent on a consensus arrived at through discussion free from domination' (Habermas, 1971, p. 6). Somewhat later he argued that 'in the last analysis it is the communicative forms of scientific and scholarly argument that hold university learning processes . . . together' and that 'the enterprise of the cooperative search for truth refers back to . . . structures of public argumentation' (Habermas, 1989a, p. 124). In a more theoretical and philosophical vein Habermas has sought to link knowledge into theories of language and action (Habermas, 1984, 1987b). One of the key purposes of this development has been to provide arguments for retaining the notion that knowledge can be emancipatory and that it is possible to raise and redeem truth claims.

The raising and redemption of truth claims

A key element in Habermas's theories of language and action has been a rejection of the argument commonly associated with postmodernism, but also pre-figured in the later arguments of Wittgenstein (see Unwin, 1992, pp. 183–4), that language functions as such an opaque medium of representation that notions of a 'true' correspondence with any world outside language is a mere deception (see Barnes and Duncan, 1992; Strohmayer and Hannah, 1992). Habermas recognises that the notion of language as a transparent mirror on extra-linguistic worlds is unrealistic (see Habermas, 1987b, 1992) but argues that this does not mean that language is unconnected with, and that people do not make statements about, these extra-linguistic worlds. Indeed Habermas argues that in the act of uttering, language and the speaker are situated within, and hence make (albeit often implicitly) 'validity' or 'truth' claims in relation to three dimensions of reality: an 'external world' of objects and events, an 'inner world' of our personal experience, and a 'social world' of expectations and obligations (Habermas, 1979, 1984, 1987b; see Unwin, 1992, p. 42 for a brief summary).

The precise status of these 'worlds' and whether all language use makes reference to them has been a subject of considerable debate (see for example McCarthy, 1982; Thompson, 1982, 1984; Culler, 1985; White, 1988; Phillips, forthcoming b). There is certainly some suspicion that the formalism and philosophical style of Habermas's analysis of language is unhelpful, that he underplays the extent to which there is 'intertextuality' (Barnes and Duncan, 1992, p. 7) in language use and that he overplays the role of validity claims in language use. On the other hand, it can be suggested that Habermas's argument that language has extra-linguistic referents is of value, not least in understanding academic discourse insofar as this discourse is recognised as being, at least in part, 'about making social life intelligible' (Gregory, 1991, p. 18). As Crang (1992, p. 534) argues, while writing human geography should be recognised as being 'creative and imaginative', it does generally not purport to be 'purely imaginary', but rather 'claims to represent living human beings'. As such it is an interpretation of something: indeed if one follows Habermas's arguments it will be an interpretation which is staking claims on how things are, how they ought to be and who one is.

It is important to note that by suggesting that statements contain validity claims, Habermas is not suggesting that these claims are necessarily 'valid', merely that they are making claims to be considered valid. As Thompson

(1990) puts it, when we put forward an 'interpretation', a statement of how we think things are, we

> make a claim which could, we suppose, be defended or sustained in some way. We do not necessarily suppose that our interpretation is the only possible or only sensible interpretation, but we do suppose that it is a justifiable one, one that *could* be justified if we were called upon to do so. (Thompson, 1990, p. 320)

Thompson adds that making social interpretations is 'inherently risky, conflict-laden and open to dispute' because it 'makes a claim about something which may differ from other views, including the views of those who make up the social world and whose everyday understanding may be the object of interpretation' (Thompson, 1990, p. 320).

The notion that academic discourse involves making claims which may well be in conflict with the views of social subjects raises clear ethical, moral and political dilemmas, particularly when the dark side of many enlightenment projects – interpretations advanced in the name or interests of groups seen to be subject to domination – has been so widely documented. However, for Habermas raising validity claims, however problematic it may become, simply cannot be avoided by any that use language. What is of significance for Habermas is that some people, particularly postmodernists, appear to wish to deny or suspend this aspect of language use. Habermas (1987a, 1989a) argues that a common feature within the work of Heidegger, Foucault, Lyotard and Derrida is a desire to separate their writings from notions of truth and falsity, right and wrong, emancipation and domination: yet in their writings they smuggle in such notions unannounced. Habermas suggests that postmodernists are guilty of 'performative contradiction': they fail to enact their own arguments fully. These performative contradictions are not, Habermas suggests, intentional aims but reflect the impossibility of escaping the presentation of truth claims in language use.

Habermas's criticism of postmodernists as enacting performative contradictions have been elaborated further. Nancy Fraser, for example, has suggested that despite the enormous value of Foucault's work, both academically and politically, his work ends up 'inviting questions that it is structurally unequipped to answer' (Fraser, 1989, p. 27). Specifically, she argues that Foucault's work falls down in its recognition of its own normative content. Foucault suggests that his 'genealogical' approach suspends, or 'brackets', the judgemental notions of truth and falsity, right and wrong. Fraser focuses her criticism on the latter bracketing and suggests that Foucault both fails to live up to his own arguments, for instance by invoking the humanistic ideals of 'autonomy, reciprocity, dignity and human rights' (Fraser, 1989, p. 57), and also, more fundamentally,

> because he misunderstands, at least when it comes to his *own* situation, the way that norms function in social description. He assumes that he can purge all traces of liberalism from his account of modern power simply by forswearing explicit reference to the tip-of-the-iceberg notions of

> legitimacy and illegitimacy. He assumes, in other words, that these norms can be neatly isolated and excised from the larger cultural and linguistic matrix in which they are situated. He fails to appreciate the degree to which the normative is embedded in and infused throughout the whole of language at *every* level and the degree to which, despite himself, his own critique has to make use of modes of description, interpretation and judgement formed within the modern Western normative tradition.
>
> (Fraser, 1989, pp. 30–1)

McCarthy (1991) suggests that similar problems lie within the work of Derrida:

> Derrida appears to want it both ways: to undermine all logocentric concepts and yet to continue to use them for his own purposes. . . . [His] deconstruction of our logocentric culture has nothing substantive to say – at least without the ironic reminder that he couldn't possibly have meant it . . . the effect of this strategy . . . are largely sceptical. . . . To play the sceptic in ethical-political matters is, of course to adopt an ethicopolitical stance. . . . If Western reason is in the end nothing more than subjectification and objectification in the service of domination, then some form of . . . deconstruction, seems a more appropriate response than any form of . . . reconstruction. But this argument presupposes that a particular interpretation of Western history, culture, society, politics, technology, etc. – and a very global and undifferentiated one at that – is the correct one, or at least that it is superior to competing interpretations. . . . Derrida has in effect deprived himself of the means he needs to enter into that debate. . . . [Furthermore] undercutting the appeal to reason, truth and justice as presently 'coded', without offering alternatives, may harbor [*sic*] not so much the 'promise of a better world' as the 'danger' of some 'monstrous mutation'.
>
> (McCarthy, 1991, pp. 108–12)

While such comments may appear far removed from debates within much of existing rural research, one can still witness attempts to deny the presentation of truth claims. For example, it has recently been suggested that the aim of rural research should be to describe people's actions and representations 'without fear or favour' (Lowe *et al.*, 1992; see also Grove-Hills, Munton and Murdoch, 1990). This argument can be questioned at both a practical level – it ignores the extent to which researchers unconsciously determine the form of responses from those they encounter in the research process (see McDowell, 1992a, 1992b; Pile, 1991b) – and at the more theoretical level of writing meaningful interpretation. The notion of writing without fear or favour implies that one can suspend judgement. Habermas has throughout his work been highly critical of the idea that you can avoid making judgement. Indeed one can see the impossibility of suspending judgement as lying at the heart of his criticisms of positivism and '*verstehen*' philosophies of the 1960s (see Habermas, 1978, 1984, 1988) and his more recent rejection of postmodernism (Habermas, 1987a, 1990). With reference to writing accounts of intentionally motivated action, Habermas (1984), for example, suggests that understanding this form of action not only involves outlining the intentions of the agent but also involves evaluating them. Specifically, he claims:

> Speech and action are the unclarified fundamental concepts to which we have recourse when we wish to elucidate, even in a preliminary way, what it is to belong to, to be an element of a socio-cultural lifeworld. . . . The social scientist basically has no other access to the lifeworld than the social-scientific layman [*sic*] does. He must already belong in a certain way to the lifeworld whose elements he wishes to describe. In order to describe them, he must understand them; in order to understand them, he must be able in principle to participate in their production; and participation presupposes that one belongs . . . this circumstance prohibits the interpreter from separating questions of meaning and questions of validity in such a way as to secure for the understanding of meaning a purely descriptive character.
>
> (Habermas, 1984, p. 110)

Habermas argues that the notion of an interpreter simply describing meanings and intentions is nonsensical – it would lead merely to a list of 'the physical substrata of utterances' (Habermas, 1984, p. 144). Hence to describe meanings is to understand – or at least think one understands – the purposes or intentions behind the production of symbolic expression. To understand intentions and purposes in turn involves, so Habermas argues, an interpreter '*bring[ing] to mind* the reasons with which a speaker would, if necessary and under suitable conditions, defend his [*sic*] validity claims' (Habermas, 1984, p. 115). In other words to understand intentionally motivated action involves thinking how the agent of this action might explain their action given the opportunity. In this moment of thought there is, so Habermas argues, a moment of judgement:

> For reasons to be sound and for them merely to be considered sound are not the same thing. . . . That is why the interpreter cannot simply look at and understand . . . reasons without at least implicitly passing judgement on them *as* reasons, that is, without taking a positive or negative position on them. The interpreter may leave some claims to validity open . . . to leave them as open problems. But reasons can be *understood* only insofar as they are taken seriously as reasons and evaluated.
>
> (Habermas, 1990, p. 30)

These arguments of Habermas are both quite dense and complicated. Fortunately, Sayer has made the same point rather more directly by considering what should be incorporated with an interpretation of society in South Africa. He argues that any account of South Africa necessarily involves passing judgement upon apartheid and the notions of racial superiority and inferiority that inform it. Even the researcher who seeks to report the views of advocates of apartheid without 'fear or favour' is biasing their account by closing off the argument that these views are false. This study is effectively establishing a claim that there could be no other situation within South Africa apart from apartheid. As Sayer put it: 'We simply can't refuse to make *any* evaluation, negative or positive, because unless we decide whether the actors own explanations of their actions are right, we cannot decide what explanation to choose ourselves' (Sayer, 1992, p. 40). Overall, one can suggest that making judgements and positing validity claims is an inescapable aspect of writing accounts of something: this is as 'true' for the rural studies as any other academic discipline (indeed see Pile, 1990a, and Smith, 1993, for very similar arguments based on research in two rural contexts).

Knowledge involves the putting forward of validity claims: yet while this argument can be seen as a mark of distinction between Habermas and the claims of the philosophers of postmodernism, it is in a sense only a starting point within his work. Two further questions exercise Habermas rather more than that of whether or not truth claims are put forward in language: namely how is it that some claims are accepted and others rejected (McCarthy, 1978, p. 65) and what are the social consequences of accepting specific claims? Habermas suggests that validity claims are often, and more significantly, perhaps, should be, redeemed through argumentative procedures. The promissory argument has clear connections with the utopian claims of Habermas's concepts such as the public sphere and the ideal speech situation. His more existential claim is that validity claims are being redeemed all the time in the everyday practices of human life: 'the communicative practice of everyday life points to the practice of argumentation as a court of appeal . . . in which participants thematize contested validity claims and attempt to vindicate or criticize them' (Habermas, 1984, p. 18). Habermas argues that the use of language in everyday conversation presupposes an ability to establish or re-establish validity claims through argumentative procedures in which reasons are advanced, debated and evaluated. In everyday speech people ask questions such as 'what did you mean?', 'is that right?', 'did it really happen like that?', 'do you really believe what you are saying?' In other words they are questioning the validity claims of intelligibility, rightness, truthfulness and sincerity (see Phillips, 1991b, p. 38). Yet communication frequently does not stop then – the person questioned responds with clarification and justifications. In this manner validity claims become accepted or rejected through reasoned argument.

However, while such 'communicative rationality' (White, 1988) is an everyday feature of social life, it is also, Habermas argues, a feature which has been poorly recognised by the philosophers of knowledge. Indeed for Habermas the notion of communicative rationality has been systematically 'foreshortened' and 'distorted' (Habermas, 1987a, p. 311) by both positivism and post-modernism, a situation which has led to a denial of the emancipatory potential of knowledge.

The difference that knowledge can make

One of the central roles that 'critical theory' has played in geography has been as a tool to challenge 'scientistic' interpretations of scientific activity: that is, interpretations which see science as an activity above society – an activity which is conducted without political, moral or social commitment. Within geography Mercer (1984), drawing on the work of the Critical Theorist Marcuse, has argued that many geographers fall prey to a 'technocratic ideology' in that they become 'preoccupied with superficial "technical" problems and are failing to address themselves to what should be the central moral and intellectual questions of our day: "How to live, and what to do?"' (Mercer, 1984, p. 162). Mercer argued that such an ideology '*trivialises*' the search for knowledge in that knowledge becomes just the product of a technique, produced by a 'scientist' because it is their job. Knowledge is not linked to people or to the society in which it is produced: the activity of the scientist is seen to be an activity which is isolated from all other aspects of social life. Indeed, philosophies of science such as logical positivism make great play on the detachment of science from society: science is seen to be 'objective' when it is divorced from social and political influence. The scientist studies things, 'objects', without regard to their practical significance: that is without providing an answer to the questions, 'what and whom are we doing science for?' The best scientist is seen to be one who is unswayed by personal values or political direction. However, Critical Theorists such as Horkheimer, Marcuse and Habermas have argued that the scientistic interpretation of science as being above society and social interests is wrong. Scientism does not remove values and interests from the practice of science: instead it recognises only a particular and very narrow set of values related to prediction and control and denigrates all other values and interests by suggesting that they are irrational (see Habermas, 1978). In the process positivism becomes socially conservative: it becomes a 'copy theory of truth' (Hahermas, 1978, p. 69) which recognises only 'what is' and ignores 'what could be' (Sayer, 1992, p. 256).

Radical and critical geographers have long identified such conservatism: Harvey's (1973) notions of 'status quo' and 'counter-revolutionary' theories still stand as a clear critique of scientistic understandings of knowledge, while Gregory (1978), and more recently Unwin (1992), have used Habermas's early work (principally Habermas, 1978) to achieve the same ends. While one may cautiously agree with the claim that much of human geography is now 'post-positivist' (Cloke, Philo and Sadler, 1991, p. viii) – although arguably rural geography is rather less 'post-positivist' than many areas of human geography (see Cloke, 1989) – it may, if one follows some of the recent arguments of Habermas, not be 'post-conservative'. Specifically Habermas (1987a, 1989a) has argued that the philosophies associated with the term postmodernism are a new form of 'conservativism'. Amongst his arguments are the claims that postmodernists act, in a manner quite similar to the 'scientism' of positivists, to separate academia from society through the establishment of 'encapsulated expert cultures' which have 'little to offer everyday life' and that postmodernists establish a form of conservativism in that they effectively deny the possibility that conscious directed social change might lead to a reduction in the degree of domination and repression that people face.

These arguments have been echoed in recent discussion amongst geographers. Rogers (1992), for example, has noted that the arguments of post-modernists, poststructuralists, postcolonialists and feminists have all provided many reasons for giving up 'global claims' including political arguments based around notions of common humanity and emancipation, and yet, personally, he would 'not want to be placed in a position of silence on the current situation in Armenia, or for that matter Sudan, Indonesia or Brazil' (Rogers, 1992, p. 522). Duncan and Barnes, whilst championing postmodernism as an 'inescapably radical and relativistic epistemology', note that the crisis of representation on which this epistemology dwells is a crisis for only 'a tiny coterie of "hyper"-educated intellectuals' (Duncan and Barnes, 1992, p. 251). Even some of the most renowned exponents of postmodernism in geography, such as Dear, Gregory and Soja, have drawn back from adopting its full implications, not least because they fear its implication for a 'politically progressive' geography. Habermas, however, makes this point even more forcibly:

> praise of multiplicity, apology for the contingent and the private, celebration of rupture, difference, and the moment, the revolt of the marginal areas against the centres, the mustering of the extraordinary against triviality – all this should not become an escape from problems that can be solved . . . by daylight . . . cooperatively . . . with the last drops of solidarity that is almost completely drained of its life blood.
>
> (Habermas, 1989a, p. 204)

Recognition of a problem does not, however, mean that a solution is readily to hand. Habermas may provide us with some pointers – he has after all been pointing out the 'conservatism' of postmodernism for several years now. Amongst the pointers is his analysis of language, knowledge and society. From his critical theory one can learn that academics are as 'socially constructed' as any of those that they study; that academics, including postmodern ones, put forward claims about how the world is and could be structured; and that academics are participants in the exercise of, and in resistance to, relations of domination. This is not to say that Habermas's critical theory is above criticism. Amongst the lessons one can learn from his critics are that even actions motivated by the most 'emancipatory' of ideals may establish relations of domination; that academics do not occupy central 'steering positions' in society; nor are they the only producers of knowledge. However, these criticisms do not mean that what academics do is completely external to society; nor that, in at least some respects and in some areas of social life, 'knowledge' (albeit partial and highly likely to be subject to revision) of the predicaments of ourselves and others cannot lead us to make the decisions that will lead towards a better future. Whilst many rural researchers may see the notion of utopian thought – by which Habermas means ideas which 'have the function of opening up alternatives for action and margins of possibility that push beyond historical contingencies' (Habermas, 1989, p. 49) – with some alarm and while advocates of postmodernism may view it as simply a strategy of domination, there may be some who will seek to build on from Habermas's claim that it is possible to outline 'the general conditions for the communicative practice of everyday life' in such a manner as to legitimate the project of establishing 'procedure[s] of discursive will-formation that would put participants *themselves* in positions to realize concrete possibilities for a better and less threatened life, on *their* own initiative and in accordance with *their own* needs and insights' (Habermas, 1989, p. 69).

SOME CONCLUDING REMARKS

This chapter has outlined some aspects of Habermas's work and has related them to the study of rural societies. Particular emphasis has been placed here on exploring the implications of his concepts of 'public' and 'private' spheres and his arguments over language, knowledge and power. It has been suggested that his concepts of public and private spheres may, albeit with some reformulation, provide useful concepts with which to write interpretations of the countryside. It has also been argued that Habermas's work is of importance in that he addresses the issue of why people should be writing about the rural: to enable people to achieve emancipation from unnecessary social constraints. It should, however, be emphasised that the arguments of Habermas are so broad that even this fairly extensive discussion has only touched on parts of his work. Many other aspects of his work may be of value to rural studies: indeed, some use has already been made of his ideas over the nature and limits of state power in late capitalist societies (see Pile, 1990a, 1990b, 1991a). Furthermore, it should noted that this discussion has extracted ideas presented by Habermas with some ten or more books written over some three decades. Within these books and in associated articles Habermas has expanded, reformulated and rejected many of his arguments and has in his two-volume *Theory of Communicative Action* sought to establish a systematic presentation of many of them. However, even this two-volume enterprise has not been able to encompass all his ideas, nor resolve all of the tensions between some of their elements, nor convince many of his critics. Habermas's elaboration of critical theory is a continuing one. This chapter has merely sought to indicate how at least some of Habermas's arguments might have some significance to how and why one attempts to 'write on the rural'.

ACKNOWLEDGEMENTS

Particular thanks are due to Chris Philo, Steve Pile and Tim Unwin for reading and commenting on an earlier draft of this chapter. While I have not accommodated all their suggestions here their comments have served to push my thoughts in some new directions. Grateful thanks also to Paul Cloke and Kate Phillips for their encouraging words during the period in which I have been exploring these ideas.

REFERENCES

Adorno, T. (1974) *Minima Moralia: Reflections from Damaged Life*, New Left Books, London.

Adorno, T. and Horkheimer, M. (1979) *The Dialectic of Enlightenment*, Verso, London.
Baker, K. M. (1992) Defining the public sphere in eighteenth century France: variations on a theme by Habermas, in C. Calhoun (ed.) *Habermas and the Public Sphere*, MIT Press, London.
Barnes, T. J. and Duncan, J. S. (1992) Introduction: writing worlds, in T. J. Barnes and J. S. Duncan (eds.) *Writing Worlds: Discourse, Text and Metaphor in the Representation of Landscape*, Routledge, London.
Bauman, Z. (1987) *Legislators and Interpreters: on Modernity, Postmodernity and Intellectuals*, Polity, Cambridge.
Bell, C. and Newby, H. (1973) *Community Studies*, Allen & Unwin, London.
Benhabib, S. (1992) *Situating the Self: Gender, Community and Postmodernism in Contemporary Ethics*, Polity, Cambridge.
Bourdieu, P. (1984) *Distinction: A Social Critique of the Judgement of Taste* Routledge & Kegan Paul, London.
Bourdieu, P. (1988) *Homo Academicus*, Polity, Cambridge.
Bradley, T. (1981) Capitalism and countryside: rural sociology as political economy, *International Journal of Urban and Regional Research*, Vol. 5, pp. 581–7.
Burgess, J. (1990) The production and consumption of environmental meanings in the mass media: a research agenda for the 1990s, *Transactions, Institute of British Geographers* (new series), Vol. 15, no. 2, pp. 139–61.
Carr, S. and Tait, J. (1990) Difference in the attitudes of farmers and conservationists and their implications, *Journal of Environmental Management*, Vol. 32, pp. 281–94.
Carr, S. and Tait, J. (1991) Farmers' attitudes to conservation, *Built Environment*, Vol. 16, no. 3, pp. 218–31.
Cloke, P. (1989) Rural geography and political economy, in R. Peet and N. Thrift (eds.) *New Models in Geography*, Volume 1, Unwin Hyman, London.
Cloke, P. and Davies, L. (1992) Deprivation and lifestyles in rural Wales. I: Towards a cultural dimension, *Journal of Rural Studies*, Vol. 8, no. 4, pp. 349–58.
Cloke, P. and Little, J. (1990) *The Rural State? Limits to Planning in Rural Society*, Oxford University Press.
Cloke, P. and Milbourne, P. (1992) Deprivation and lifestyles in rural Wales. II: Rurality and the cultural dimension, *Journal of Rural Studies*, Vol. 8, no. 4, pp. 359–71.
Cloke, P. and Phillips, M. (1991) *Social Recomposition and Class Relations: Issues in the Design of a Questionnaire*. ESRC Middle Class Project Working Paper No. 3, St David's University College, Lampeter.
Cloke, P., Philo, C. and Sadler, D. (1991) *Approaching Human Geography: An Introduction to Contemporary Theoretical Debates*, Paul Chapman, London.
Cloke, P. and Thrift, N. (1990) Class and change in rural Britain, in T. Marsden, P. Lowe and S. Whatmore (eds.) *Rural Restructuring: Global Processes and Local Responses*, Fulton, London.
Cohen, A. (ed.) (1982) *Belonging: Identity and Social Organisation in British Rural Cultures*, Manchester University Press.
Cohen, A. (ed.) (1986) *Symbolising Boundaries: Identity and Diversity in*

British Cultures, Manchester University Press.

Corrigan, P. and Sayer, D. (1985) *The Great Arch: English State Formation as Cultural Revolution*, Blackwell, Oxford.

Cox, G. and Lowe, P. (1984) Agricultural corporatism and conservation politics, in T. Bradley and P. Lowe (eds.) *Locality and Rurality: Economy and Society in Rural Regions*, Geo Books, Norwich.

Crang, P. (1992) The politics of polyphony: reconfigurations in geographical authority, *Environment and Planning D: Society and Space*, Vol. 10, no. 5, pp. 527–49.

Culler, J. (1985) Communicative competence and normative force, *New German Critique*, Vol. 35, pp. 133–44.

Delphy, C. and Leonard, D. (1986) Class analysis, gender analysis and the family, in R. Crompton and M. Mann (eds.) *Gender and Stratification*, Polity, Cambridge.

Dews, P. (ed.) (1986) *Autonomy and Solidarity: Interviews with Jürgen Habermas*, Verso, London.

Doel, M. (1992) In stalling deconstruction: striking out the postmodern, *Environment and Planning D: Society and Space*, Vol. 10, no. 2, pp. 163–80.

Doel, M. (1993) Proverbs for paranoids: writing geography on hollowed ground, *Transactions, Institute of British Geographers*, Vol. 18, no. 3, pp. 377–94.

Duncan, J. S. and Barnes, T. J. (1992) Afterword, in T. J. Barnes and J. S. Duncan (eds.) *Writing Worlds: Discourse, Text and Metaphor in the Representation of Landscape*, Routledge, London.

Eley, G. (1992) Nations, publics and political cultures: placing Habermas in the nineteenth century, in C. Calhoun (ed.) *Habermas and the Public Sphere*, MIT Press, London.

Flynn, A., Lowe, P. and Cox, G. (1990) *The Rural Land Development Process*. ESRC Countryside Change Initiative Working Paper 6, University of Newcastle.

Foucault, M. (1980) *Power/Knowledge: Selected Interviews and Other Writings 1972–1977*, Harvester, Brighton.

Fraser, N. (1989) *Unruly Practices: Power, Discourse and Gender in Contemporary Social Theory*, Polity, Cambridge.

Friedmann, H. (1986) Patriarchal commodity production, *Social Analysis*, Vol. 20, pp. 47–55.

Goodwin, M. (1989) Uneven development and civil society in Western and Eastern Europe, *Geoforum*, Vol. 20, no. 2, pp. 151–9.

Gorz, A. (1982) *Farewell to the Working Class*, Pluto, London.

Gregory, D. (1978) *Ideology, Science and Human Geography*, Hutchinson, London.

Gregory, D. (1989) Areal differentiation and post-modern human geography, in D. Gregory and R. Walford (eds.) *Horizons in Human Geography*, Macmillan, London.

Gregory, D. (1991) Interventions in the historical geography of modernity: social theory, spatiality and the politics of representation, *Geografiska Annaler*, Vol. 73B, no. 1, pp. 17–44.

Gregory, D. (1992) Interventions in the historical geography of modernity: social theory, spatiality and the politics of representation, *Geografiska Annaler*, Vol. 73B, pp. 12–44.
Grove-Hills, J., Munton, R. and Murdoch, J. (1990) *The Rural Land Development Process: Evolving a Methodology*. ESRC Countryside Change Initiative Working Paper 8, University of Newcastle.
Habermas, J. (1971) *Towards a Rational Society: Student Protests, Science, and Politics*, Heinemann, London.
Habermas, J. (1974) *Theory and Practice*, Heinemann, London.
Habermas, J. (1978) *Knowledge and Human Interests* (2nd edn), Heinemann, London.
Habermas, J. (1979) *Communication and the Evolution of Society*, Heinemann, London.
Habermas, J. (1982) A reply to my critics, in J. Thompson and D. Held (eds.) *Habermas: Critical Debates*, Macmillan, London.
Habermas, J. (1983) *Philosophical-Political Profiles*, Polity, Cambridge.
Habermas, J. (1984) *The Theory of Communicative Action, Volume 1: Reason and the Rationalization of Society*, Heinemann, London.
Habermas, J. (1987a) *The Philosophical Discourse of Modernity: Twelve Lectures*, Polity, Cambridge.
Habermas, J. (1987b) *The Theory of Communicative Action, Volume 2: The Critique of Functionalist Reason*, Polity, Cambridge.
Habermas, J. (1988) *On the Logic of the Social Sciences*, Polity, Cambridge.
Habermas, J. (1989a) *The New Conservatism: Cultural Criticism and the Historians' Debate*, Polity, Cambridge.
Habermas, J. (1989b) *The Structural Transformation of the Public Sphere: An Inquiry into a Category of Bourgeois Society*, Polity, Cambridge.
Habermas, J. (1990) *Moral Consciousness and Communicative Action*, Polity, Cambridge.
Habermas, J. (1991) A reply, in A. Honneth and H. Joas (eds.) *Communicative Action: Essays of Jürgen Habermas's 'The Theory of Communicative Action'*, MIT, Cambridge, Massachusetts.
Habermas, J. (1992) *Postmetaphysical Thinking: Philosophical Essays*, Polity, Cambridge.
Habermas, J. (1993) *Justification and Application: Remarks on Discourse Ethics*, Polity, Cambridge.
Halfacree, K. (1993) Locality and social representation: space, discourse and alternative definitions of the rural, *Journal of Rural Studies*, Vol. 9, no. 2, pp. 23–37.
Harper, S. (1987) A humanistic approach to the study of rural populations, *Journal of Rural Studies*, Vol. 3, pp. 309–19.
Harper, S. (1989) The British rural community: an overview of perspectives, *Journal of Rural Studies*, Vol. 5, no. 2, pp. 161–84.
Harvey, D. (1973) *Social Justice and the City*, Edward Arnold, London.
Hobsbawm, E. and Rudé, G. (1969) *Captain Swing*, Lawrence & Wishart, London.
Howkins, A. (1991) *Reshaping Rural England: A Social History, 1850–1925*, HarperCollins Academic, London.

Hoggart, K. (1988) Not a definition of rural, *Area*, Vol. 20, pp. 35–40.
Hoggart, K. (1990) Let's do away with rural, *Journal of Rural Studies*, Vol. 6, pp. 245–57.
Hoggart, K. and Buller, H. (1987) *Rural Development: A Geographical Perspective*, Croom Helm, London.
Horkheimer, M. (1937) Traditionelle und Kritische Theories, *Zeitschrift für Sozialforschung*, Vol. 6, no. 2; translated into English as Traditional and critical theory, in M. Horkheimer (1973) *Critical Theory: Selected Essays*, Seabury Press, New York.
Howell, P. (1993) Public space and the public sphere: political theory and the historical geography of modernity, *Environment and Planning D: Society and Space*, Vol. 11, pp. 303–22.
Landes, J. (1988) *Women and the Public Sphere in the Age of the French Revolution*, Cornell University Press, Ithaca.
Little, J. (1987) Gender relations in rural areas: the importance of women's domestic role, *Journal of Rural Studies*, Vol. 3, no. 4, pp. 335–42.
Littlejohn, J. (1963) *Westrigg: the Sociology of a Cheviot Parish*, Routledge & Kegan Paul, London.
Livingstone, D. (1992) *The Geographical Tradition: Episodes in the History of a Contested Enterprise*, Blackwell, Oxford.
Lowe, P. and Bodiguel, M. (eds.) (1990) *Rural Studies in Britain and France*, Bellhaven, London.
Lowe, P., Marsden, T. and Munton, R. (1990) *The Social and Economic Restructuring of Rural Britain: A Position Statement*. ESRC Countryside Change Initiative Working Paper 2, University of Newcastle.
Lowe, P., Marsden, T., Murdoch, J., Flynn, A. and Munton, R. (1992) The social and economic restructuring of rural Britain: some results and research implications from the countryside change initiative. Paper given at the annual meeting of the Rural Economy and Society Study Group, Hull.
MacKenzie, S. (1989) Women in the city, in R. Peet and N. Thrift, (eds.) *New Models in Geography, The Political-Economy Perspective* (Vol. 2), Unwin Hyman, London.
Mahar, C. (1991) On the moral economy of country life, *Journal of Rural Studies*, Vol. 7, no. 4, pp. 363–72.
Marcuse, H. (1964) *One Dimensional Man*, Routledge & Kegan Paul, London.
Marsden, T. (1988) Exploring political economy approaches in agriculture, *Area*, Vol. 20, pp. 315–22.
Marsden, T. (1992) Exploring a rural sociology for the Fordist transition: incorporating social relations into economic restructuring, *Sociologia Ruralis*, Vol. XXXII, no. 2/3, pp. 209–30.
Marsden, T. and Murdoch, J. (1990) *Restructuring Rurality: Key Areas for Development in Assessing Rural Change*, ESRC Countryside Change Initiative Working Paper no. 4, University of Newcastle.
Marsden, T., Whatmore, S., Munton, R. and Little, J. (1986) Toward a political economy of capitalist agriculture: a British perspective, *International Journal of Urban and Regional Research*, Vol. 10, pp. 498–521.
Marshall, G., Rose, D., Newby, H. and Vogler, C. (1988) *Social Class in Modern Britain*, Hutchinson, London.

Matless, D. (1992) An occasion for geography: landscape, representation and Foucault's corpus, *Environment and Planning D: Society and Space*, Vol. 10, no. 1, pp. 41–56.

McCarthy, T. (1978) *The Critical Theory of Jürgen Habermas*, Polity, Cambridge.

McCarthy, T. (1982) Rationality and relativism: Habermas's 'overcoming' of hermeneutics, in J. Thompson and D. Held (eds.) *Habermas: Critical Debates*, Macmillan, London.

McCarthy, T. (1989) Introduction, in J. Habermas (ed.) *The Structural Transformation of the Public Sphere: An Inquiry into a Category of Bourgeois Society*, Polity, Cambridge.

McCarthy, T. (1991) *Ideals and Illusions: On Reconstruction and Deconstruction in Contemporary Critical Theory*, MIT Press, Cambridge, Massachusetts.

McDowell, L. (1983) Towards an understanding of the gender division of urban space, *Environment and Planning D: Society and Space*, Vol. 1, pp. 59–72.

McDowell, L. (1992a) Doing gender: feminism, feminists and research methods in human geography, *Transactions, Institute of British Geographers* (new series), Vol. 17, no. 4, pp. 399–416.

McDowell, L. (1992b) Valid games? A response to Erica Schoenberger, *Professional Geographer*, Vol. 44, pp. 212–15.

Mercer, D. (1984) Unmasking Technocratic Geography, in M. Billinge, D. Gregory and R. Martin (eds.) *Recollections of a Revolution: Geography as Spatial Science*, Macmillan, London, pp. 153–9.

Middleton, A. (1984) Marking boundaries: men's space and women's space in a Yorkshire village, in T. Bradley and P. Lowe (eds.) *Locality and Rurality: Economy and Society in Rural Regions*, Geo Books, Norwich.

Mormont, M. (1990) Who is rural? or, How to be rural: towards a sociology of the rural, in T. Marsden, P. Lowe and S. Whatmore (eds.) *Rural Restructuring: Global Processes and Their Responses*, Fulton, London.

Murdoch, J. and Marsden, T. (1990) *Reconstituting the Rural in an Urban Region: New Villages for Old?* ESRC Countryside Change Initiative Working Paper 26, University of Newcastle.

Murdoch, J. and Pratt, A. (1993) Rural Studies: modernism, postmodernism and the post-rural, *Journal of Rural Studies*, Vol. 9, no. 4, pp. 411–27.

Newby, H. (1977) *The Deferential Worker*, Allen Lane, London.

Newby, H., Bell, C., Rose, D. and Saunders, P. (1978) *Property, Paternalism and Power: Class and Control in Rural England*, Hutchinson, London.

Newby, H. and Buttel, F. H. (1980) Toward a critical rural sociology, in F. H. Buttel and H. Newby (eds.) *The Rural Sociology of the Advanced Societies: Critical Perspectives*, Croom Helm, London, pp. 1–35.

Newby, H., Bujira, J., Littlewood, P., Rees, G. and Rees, T. (eds.) (1985) *Restructuring Capital: Recession and Reorganization in Industrial Society*, Macmillan, Basingstoke.

O'Connor, J. (1987) *The Meaning of Crisis: a Theoretical Introduction*, Basil Blackwell, Oxford.

Ogborn, M. (1993) Public sphere/public space: the politics of the city in late eighteenth century London. Paper presented at annual conference, Institute of British Geographers, Egham.

Pateman, C. (1983) Feminist criticisms of the public/private dichotomy, in I. Benn and S. Gauss (eds.) *Public and Private in Social life*, Croom Helm, Beckenham.

Phillips, M. (1991a) *Analysing Middle Class Relations.* ESRC Middle Class Project Working Paper No. 2, St David's University College, Lampeter.

Phillips, M. (1991b) Market exchange and social relations: the practices of food circulation in and to the Three Towns of Plymouth, Devonport and Stonehouse, circa 1800–1870. Unpublished Ph.D. thesis, University of Exeter.

Phillips, M. (1993) Rural gentrification and the processes of class colonisation, *Journal of Rural Studies*, Vol. 9, no. 2, pp. 123–4.

Phillips, M. (1994) Teaching and learning philosophy in geography from textbooks, *Journal of Geography in Higher Eductation*, Vol. 18, no. 1, pp. 114–22.

Phillips, M. (forthcoming a) Economic and communicative integration in market exchange: illustrations from a study of food circulation in nineteenth century Britain, submitted to *Journal of Historical Geography.*

Phillips, M. (forthcoming b) Habermas, critical theory and human geography, submitted to *Environment and Planning D: Society and Space.*

Philo, C. (1992) Neglected rural geographies: a review, *Journal of Rural Studies*, Vol. 8, no. 2, pp. 193–207.

Philo, C. (1993) Postmodern rural geography? A reply to Murdoch and Pratt, *Journal of Rural Studies*, no. 9, pp. 429–436.

Pile, S. (1990a) Depth hermeneutics and critical human geography, *Environment and Planning D: Society and Space*, Vol. 8, no. 2, pp. 211–32.

Pile, S. (1990b) *The Private Farmer: Transformation and Legitimation in Advanced Capitalist Agriculture*, Dartmouth Press, Aldershot.

Pile, S. (1991a) 'A load of bloody idiots': Somerset dairy farmers' views of their political world, *Political Geography Quarterly*, Vol. 10, no. 4, pp. 405–21.

Pile, S. (1991b) Practising interpretative geography, *Transactions, Institute of British Geographers*, Vol. 16, pp. 458–69.

Pile, S. and Rose, G. (1992) All or nothing? Politics and critique in the modernism–postmodernism debate, *Environment and Planning D: Society and Space*, Vol. 10, no. 2, pp. 123–36.

Rawls, J. (1980) Kantian constructivism in moral theory: the Dewey Memorial Lectures 1980. *Journal of Philosophy*, Vol. 77, pp. 515–72.

Redclift, N. (1985) The contested domain: gender, accumulation and the labour process, in N. Redclift and E. Mingione (eds.) *Beyond Employment: Household, Gender and Subsistence*, Basil Blackwell, Oxford.

Rees, A. (1950) *Life in a Welsh Countryside*, University of Wales Press, Cardiff.

Rees, G. (1984) Rural regions in National and International economies, in T. Bradley and P. Lowe (eds.) *Locality and Rurality: Economy and Society in Rural Regions*, Geobooks, Norwich.

Robinson, G. (1990) *Conflict and Change in the Countryside*, Bellhaven, London.

Roderick, R. (1986) *Habermas and the Foundations of Critical Theory*, Macmillan, London.
Rogers, A. (1992) The boundaries of reason: the world, the homeland and Edward Said, *Environment and Planning D: Society and Space*, Vol. 10, no. 5, pp. 511–26.
Rorty, R. (1982) *Consequences of Pragmatism*, University of Minnesota Press, Minneapolis.
Rose, G. (1991) The struggle for political democracy: emancipation, gender, and geography, *Environment and Planning D: Society and Space*, Vol. 8, no. 4, pp. 395–408.
Rose, G. (1993) *Feminism and Geography: The Limits of Geographical Knowledge*, Polity, Cambridge.
Ryan, M. P. (1992) Gender and public access: women's politics in nineteenth century America, in C. Calhoun (ed.) *Habermas and the Public Sphere*, MIT Press, London.
Sayer, A. (1992) *Method in Social Science* (2nd edn), Routledge, London.
Schudson, M. (1992) Was there ever a public sphere? If so, when? Reflections on the American case, in C. Calhoun (ed.) *Habermas and the Public Sphere*, MIT Press, London.
Siltanen, J. and Stanworth, M. (1984) *Women and the Public Sphere: A Critique of Sociology and Politics*, Hutchinson, London.
Slater, D. (1992) On the borders of social theory: learning from other regions, *Environment and Planning D: Society and Space*, Vol. 10, no. 3, pp. 307–27.
Smith, S. J. (1993) Bounding the Borders: claiming space and making place in rural Scotland, *Transactions, Institute of British Geographers*, Vol. 18, no. 3, pp. 291–308.
Stacey, M. (1981) The division of labour revisited or overcoming the two Adams, in P. Abrams *et al.* (eds.) *Practice and Progress: British Sociology 1950–1980*, George Allen & Unwin, London.
Stallybrass, P. and White, A. (1986) *The Politics and Poetics of Transgression*, Methuen, London.
Strohmayer, U. (1993) Beyond theory: the cumbersome materiality of shock, *Environment and Planning D: Society and Space*, Vol. 11, pp. 323–47.
Strohmayer, U. and Hannah, M. (1992) Domesticating postmodernism, *Antipode*, Vol. 24, pp. 29–55.
Taylor, C. (1991) Language and society, in A. Honneth and H. Joas (eds.) *Communicative Action: Essays on Jürgen Habermas's 'The Theory of Communicative Action'*, MIT Press, Cambridge, Massachusetts.
Thompson, J. B. (1982) Universal pragmatics, in J. Thompson and D. Held (eds.) *Habermas: Critical Debates*, Macmillan, London.
Thompson, J. B. (1984) Universal pragmatics: Habermas's proposals for the analysis of language and truth, in J. B. Thompson (ed.) *Studies in the Theory of Ideology*, Polity, Cambridge.
Thompson, J. B. (1990) *Ideology and Modern Culture*, Polity, Cambridge.
Unwin, T. (1992) *The Place of Geography*, Longman, Harlow.
Urry, J. (1981) *The Anatomy of Capitalist Society*, Macmillan, London.
Urry, J. (1984) Capitalist restructuring, recomposition and the regions, in T.

Bradley and P. Lowe (eds.) *Locality and Rurality: Economy and Society in Rural Regions*, Geobooks, Norwich.
Warner, M. (1992) The mass public and the mass subject, in C. Calhoun (ed.) *Habermas and the Public Sphere*. MIT Press, London.
Whatmore, S. (1991) *Farming Women: Gender, Work and Family Enterprise*, Macmillan, London.
Whatmore, S., Munton, R. and Marsden, T. (1992) The rural restructuring process: emerging divisions of agricultural property rights, *Regional Studies*, Vol. 24, no. 3, pp. 235–45.
White, S. K. (1988) *The Recent Work of Jürgen Habermas: Reason, Justice and Modernity*, Cambridge University Press.
Wilmot, S. (1990) *'The Business of Improvement': Agriculture and Scientific Culture in Britain, c. 1700–c. 1870*. Historical Geography Research Group: Research Series, No. 24.
Wohin, R. (1989) Introduction, in J. Habermas *The New Conservatism: Cultural Criticism and the Historians' Debate*, Polity, Cambridge.
Wright, S. (1992) Image and analysis: new directions in community studies, in B. Short (ed.) *The English Rural Community*, Cambridge University Press.
Zaret, D. (1992) Religion, science and printing in the public spheres in seventeenth century England, in C. Calhoun (ed.) *Habermas and the Public Sphere*, MIT Press, London.

Chapter 3

Something Resists: Reading–Deconstruction as Ontological Infestation (Departures from the Texts of Jacques Derrida)

Marcus Doel

> I abandon this reading to you: polysemia or even dissemination drags it far from any shore, preventing what you call an event from ever arriving. Let the net float, the infinitely tortuous play of knots and links which catches this sentence in its drawing.
>
> (Derrida, 1988, p. 25)

> This letting beyond essence, 'more passive than passivity,' hear it as the most provocative thought today.
>
> (Derrida, 1991, p. 424)

THE MOTIONLESS VOYAGE OF BECOMING INDIFFERENT

> I hate the reading idler. He who knows the reader, does nothing for the reader. Another century of readers – and spirit itself will sink. That everyone can learn to read will ruin in the long run not only writing, but thinking too.
>
> (Nietzsche, 1969, p. 148)

> I change too quickly: my today refutes my yesterday. When I ascend I often jump over steps, and no step forgives me that.
>
> (Nietzsche, 1969, p. 69)

(All of this will have entailed the quest for a line 'between two translations, one governed by the classical model of transportable univocality or of formalizable polysemia, and the other, which goes over into dissemination – this line also passes between the critical and the deconstructive' (Derrida, 1991, p. 262).

Find a text
To read
If you can).

Derrida (1982, p. x) opens his attempt 'To tympanize – philosophy' with the phrase '*Being at the limit*: these words do not yet form a proposition, and even less a discourse. But there is enough in them, provided that one plays upon it, to engender almost all of the sentences in this book.' Within this essay I will endeavour to uncover what takes place at the limit of the following phrase: the world is text. But from the outset, let us be clear on the following point: I do not wish to say that the world is like a text; it is a text (End of story).

Anthropomorphic discourses have always thought that at the limit of a text, reading begins. And so it should. But the problem which Derrida raises with the phrase 'Being at the limit' is the problem of never arriving at the limit. And if, strictly speaking, reading can only begin by broaching and then breaching the limit of the text, then all of this will have taken place in 'nonreading.' The phrase 'the world is text' does not designate a delimited space which could be the subject of a reading. '*Is* does not therefore signify *is there*, and even less so does it signify *is real. Is* doesn't signify anything. . . . Rather *is* would be: *Is it happening?* (the *it* indicating an empty place to be occupied by a referent)' (Lyotard, 1988, p. 79). 'Derive all the consequences of this: they involve each element, each term of the preceding sentence' (Derrida, 1991, p. 268). Accordingly, the phrase 'the world is text' does not signify, *pace* Barnes and Duncan (1992), here it is, read it! (Nor does it signify, *pace* Bhaskar, 1989, it is real, reclaim it!): it does not signify. To the contrary, the phrase 'the world is text' would be: Is it (but what?) happening? It is in this sense that Derrida (1988, p. 25) writes 'I abandon this reading to you'. Meaning: find a text, to read, if you can!

For the sake of conversational expediency, let's say that I accept the challenge and agree to cover, to cover *in full*, a deconstructive reading of contemporary rural change. And in order to facilitate this coverage let's agree to hold open three distinct spaces, or fields, within the duration of this essay: one for holding the things which we wish to approach and cover (working title: *Ontology*); one for developing our strategies of approach (working title: *Methodology*); and one for holding and developing our strategies of reading (working title: *Epistemology*). However, one must not forget to think the totality of these three fields within a single movement. In other words, ontology, methodology and epistemology are distinct but contiguous 'steps' in the movement of reading. Moreover, methodology, like language, is traditionally assigned a purely communicative function insofar as it denotes the 'way,' 'path' or 'road' to knowledge. In other words, methodology, like language, is usually deployed in an architectonic attempt to bridge the gap, rift or abyss between ontology and epistemology. This is why the assignment to labour – organised according to the principle of efficient and effective communication – is at the core of both dialectics and metaphysics. It should come as no surprise, therefore, to discover that the thrifty, miserly, serious and restricted economies of Habermas (1987) and Rorty (1989, 1991) have been pitted against the profligate, decadent, playful and general economies of Derrida (1981a, 1987) and Lyotard (1988). Without descending here into the detail of this disputation, suffice it to say that the need for a purely functional (and therefore re-presentational) medium will arise only when one begins with an *a priori* separation or division (which is to say, whenever one begins with a desire to step or stride).

In a gesture directed towards the totality of a single, monumental movement which strides across three fields, and in an endeavour to be in step with the propaedeutic of reading (which is also the propaedeutic of humanism and hermeneutics), let's agree to write under the cover of a mark which would gather our fields together under a common banner without effacing their specificity. This mark would need to be more than a mere name. It would need to be compact, functional, elegant and homely. A commune? Perhaps. A community? Perhaps. A village? Perhaps. An acronym? Yes, yes. OME. OME will be both the gesture which gathers and shelters our movement (the structuration of our reading) and the mark which maintains and relays the trace of our steps (the trail of footprints left imprinted upon the surface of the text). 'Gathering,' 'sheltering,' 'maintaining,' 'relaying,' 'tracing,' 'trailing' and 'imprinting:' these are the footprints which are in step with the dangerous landscape and arduous terrain of the text. The reader is always caught in step, in her own footprints, ensnaring herself on the exposed pathway (*hodos*) which striates the text. Everything she reads has traps! Everything she covers has traps!

Pathways, then, are literally trodden and imprinted into the ground. They are stamped into place and codified and ossified through endless repetition. Finally, having been established on the surface of the ground, pathways must be maintained through an interminable beating of matter under foot. Henceforth, it is easy to understand why there is no intrinsic difference between grass, ground and tarmac: all are sites made available to tread! OME is the tread of the reader. In other words, all of the steps take place in reading and, therefore, on the open landscape of the text. To this extent, reading is necessarily violence: to read with a hammer. 'Leave a trace in the text if you can' teased Bennington (1989a, p. 84). 'And the enlightened man shall learn to *build* with mountains!' proclaimed Nietzsche (1969, p. 128). But the difficulty of being in step, of being left to forge and maintain a pathway, is negated by the serenity of the monumental stride across the entire space occupied by the three field system. This stride is as simple as ONE, TWO: a *single* step from there to here, from text to reading, from world to mind, from matter to sense, from error to truth, from doubt to certainty, from sign to referent, from outside to inside, etc. What could be simpler than a single movement? What could be simpler than a passive deference to gravity? What could be simpler than a solitary glide of a metronome's pendulum? 'In the mountains the shortest route is from peak to peak' wrote Nietzsche (1969, p. 67), 'but for that you must have long legs. Aphorisms should be peaks, and those to whom they are spoken should be big and tall in stature.' And what could be more aphoristic, peaked and strident than an acronym? Indeed, what could be more thrifty, miserly, serious and restricted than an acronym? Communicative efficiency always engenders the quest for the more expedient than the expedient, and the more minimal than the minimal – incessantly, it leads from the code, through the cryptic, to the unreadable and beyond! Truly, I say to you: on this motionless voyage, 'The sedentary thinker will be left behind' (Wood, 1987, p. 148). OME – let it be your talisman, your sanctuary, your crypt. Yet one should always remember that stepping, walking, marching and running are best accomplished at speed. Whilst there is no limit to how fast and how far the step can go, there is a limit to how slow it can go. Deceleration and dawdling are potentially fatal to its success. As a general principle, be pragmatic. As a golden rule, never stop. Stopping in mid step, *between* ONE and TWO, in defiance of gravity, will only serve to open an abyss beneath you. If this happens, balance rapidly becomes impossible as momentum is lost! Such is the catastrophe which cowers beneath the curvature of gravity's rainbow (This is why logic and philosophy are very strict on the following point: access is barred to the excluded middle – 'Private Property,' 'Keep Out,' 'No Through Road,' 'Trespassers Will Be Prosecuted,' etc.).

The single, monumental, arched and passive stride of OME takes place in reading, as reading. Its aim is to occupy and hold space through the deployment of communicative functions and installations. Its desire is to cover and account for all of the ground, and to erase the resistance of asignifying matter. Finally, its strategy is to trample criss-crossing pathways into the ground. (Stamping imprints onto matter: there is no difference in kind between criss-crossing pathways and postal networks, Derrida, 1987).

> A path that mounted defiantly through boulders and rubble, a wicked, solitary path that bush or plant no longer cheered: a mountain path crunched under my foot's defiance. Striding mute over the mocking clatter of pebbles, trampling the stones that made it slip: thus my foot with effort forced itself upward. Upward – despite the spirit that drew it downward, drew it towards the abyss, the Spirit of Gravity, my devil and arch-enemy. Upward – although he sat upon me, half dwarf, half mole; crippled, crippling; pouring lead-drops into my ears, leaden thoughts into my brain. 'O Zarathustra,' he said mockingly, syllable by syllable, every stone that is thrown must – fall.
>
> (Nietzsche, 1969, pp. 176–7)

The gigantesque and almost imperceptible stride of reading, of OME, of humanism and hermeneutics, and of dialectics and metaphysics, is the insatiable Spirit of Gravity: 'through him all things are ruined' (Nietzsche, 1969, p. 68). That which is hurled, that which is always in step, must – fall. 'He who will one day teach men to fly will have moved all boundary-stones; all boundary-stones will themselves fly into the air' wrote Nietzsche (1969, p. 210). Being in step is always to be subject to the Spirit of Gravity. And writing, insofar as it has the courage and strength to fight the tiresome burden and oppressive force of the Spirit of Gravity, must struggle for 'a permanent process of disordering order' (Derrida, 1989c, p. 223). Against common sense and the single, monumental stride across the ranged peaks of OME, deconstructive writing defies the Spirit of Gravity. It steps into the single step and – stops. But, 'not only am I not sure, as I never am, of being right in taking this step, I am not sure that I see in all clarity what led me to do so' wrote Derrida (1983, p. 49). 'Perhaps because I was beginning to know only too well not indeed where I was going, but where I had not so much arrived as simply stopped.'

Having stopped, in step, between ONE and TWO, I want to suggest that all of Derrida's writing can be said to take place (or rather, take space) within the duration (or rather, within the volume) of this stoppage. And insofar as 'writing always leads to more writing, and more and still more' (Rorty, 1982, p. 94), one might say that Derrida, through his prolific writing, perhaps even carcinogenic writing (Critchley, 1990), causes a relentless swelling of the purportedly imperceptible gap, fissure or crack between ONE and TWO. In this sense, being in step with a motionless voyage necessitates a distension: 'A trembling spreads out which then makes the entire old shell crack' (Derrida, 1978, p. 260). Or again: 'recall that the *written* text of philosophy (this time in its books) overflows and cracks its meaning' (Derrida, 1982, p. xxiii). In sum, deconstructive writing works to distend all of those spaces and volumes (margins, creases, limits, folds, gaps, fissures, partitions, pathways, etc.) which have been excluded (or rather, dissimulated) in a desire to maintain the deception of the Kantian tradition, which, 'no matter how much writing it does, it does not think that philosophy *should* be written' (Rorty, 1982, p. 94). Instead of writing, they would rather show (reveal, unveil, re-present, etc.). But 'There is no "picture of the world" that "we" would "make" for ourselves' writes Lyotard (1988, p. 79). 'In this sense, there is no representation.' There's only one: radical alterity, absolute singularity (And in parentheses, a word of caution: finding our way in this kind of hostile landscape is far from simple; proprietors, so-called, are not usually accustomed to producing signposts and guide books for marginal, excluded and forgotten land – find a text to leave a trace in if you can. In this sense, our aim will be neither to read, nor to show. Rather, we will aim to write).

Simplifying to the extreme, and once again for the sake of conversational expediency, deconstructive writing, in its naive guise, hurls every construction into the air and then perpetually intervenes in an attempt to prevent them from falling back into reconstructions. It is the master of deception, of trickery and of the sleight of hand (Ellis, 1989; Tallis, 1988). However, I would say that properly speaking, this is Baudrillard's (1988, 1990a), rather than Derrida's, strategy: commencing with the Herculean effort to free the real, the social and the historical from the gravitational effects of reference (i.e. anchorage) to a concrete body (Baudrillard, 1975, 1981); and culminating in the imperceptible nudge which sends them into deep space (Baudrillard, 1983, 1987, 1990b). Beyond the 'gravitational effect which maintains bodies in an orbit of signification,' writes Baudrillard (1986, p. 18), 'once "liberated" with sufficient speed, all atoms of meaning are lost in space.' It is Baudrillard's thesis that the referential masses (however defined) have imploded into cancellating black holes, or simulacra, which are antithetical to both history and genealogy (contrast, for example, the periodisation in *Simulations*, 1983, with its adsorption in *Forget Foucault*, 1987). Moreover, deriding what I will characterise as Derrida's rigorous deconstructive *practice*, Baudrillard (1990c, p. 10) writes: 'Any movement that believes it can subvert a system by its infra-structure is naive. . . . There is no need to play being against being, or truth against truth; why become stuck undermining foundations, when a *light* manipulation of appearances will do.' For Baudrillard, even the lightest of touches and the most gentle of caresses results in friction burns. Indeed, this is why he pits his playful and irresponsible 'seduction' against what he characterizes as Derrida's all too serious and laborious 'production.' In this regard, one might venture the suggestion that Baudrillard has taken Derrida (1978, p. 256) at his word: 'Laughter alone exceeds dialectics and the dialectician; it bursts out only on the basis of an absolute renunciation of meaning, an absolute risking of death.' But ultimately I do not think that Baudrillard's strategy is tenable – he is *not comic enough* (for example, compare Kellner, 1989, and Gane, 1991a, 1991b, on Baudrillard, with Megill, 1985, and Rosen, 1987, on Derrida).

As a working hypothesis, let's agree to characterize deconstruction as a disarrangement of the *desire* to systematize, construct and order. In this sense, deconstructive writing would always seek to scatter (disseminate) everything that purported to be formed, fixed and frozen to the wind (and it is particularly adept at scattering footprints and pathways, especially by means of the short-circuit [X]). In other words, deconstruction is a kind of writing (Rorty, 1982) – but not just *any* kind of writing (Derrida, 1989b) – which frustrates and prevents a determined reading from taking place. It withholds, if you like, the elements (positions) which would be the subject of a reading. This is why Derrida often portrays his deconstructive writings as so-many failed attempts to have done with the unsettling preface and enter the body of the text 'proper' (Derrida, 1987; Llewelyn, 1986; Megill, 1985). Or again: Derrida is forever exploring the terrain of writing in the vain attempt to find the edge, limit, boundary, fold, margin, pathway, threshold or crease which would allow one access either to, or from, the text. Hereafter, 'there is nothing outside of the text' (Derrida, 1976, p. 158); therefore, there will have been no inside the text either! 'If we are to approach a text,' wrote Derrida (1991, p. 256), 'it must have an edge.' But this limit, beyond which the textuality of the text would not run, is precisely that which a text in general can never possess. What would it mean, then, to be at the limit? What would it mean to think this limit and to think *through* this limit? Thought of the 'outside,' of the 'other,' of 'exteriority' and 'alterity,' and of non-negatable 'difference.' However, this list of words could never function as 'the conceptual headings under which philosophy's border could be overflowed; the overflow is its object' (Derrida, 1982, pp. xiii–xiv). Philosophy

> has recognized, conceived, posited, declined the limit according to all possible modes; and therefore by the same token, in order better to dispose of the limit, has transgressed it. *Its own limit* had not to remain foreign to it. Therefore it has appropriated the concept for itself; it has believed that it controls the margin of its volume and that it thinks its

> other. Philosophy has always insisted upon this: thinking its other. Its other: that which limits it, and from which it derives its essence, its definition, its production.... *To insist* upon thinking *its other*.... In thinking it *as such*, in recognizing it, one misses (the) missing (of) it, which, as concerns the other, always amounts to the same.
>
> (Derrida, 1982, pp. x–xii)

In short, and very schematically, there is no limit which can withhold philosophy's subl(im)ating transgression. Paradoxically, therefore, the limit is always already de-limited, disarranged and dispersed. Can there be a limit to philosophy? And who would answer this question? And with what authority? And in whose name? If there could be a limit, who would set it in place? And who would guard it? Would it be philosophy's limit? In every direction, philosophy is incapable of locating its (own) (proper) limit. Paradoxically, then, the limit of philosophy cannot be philosophy's limit. 'By right of birth, and for one time at least, these are problems put to philosophy as problems philosophy cannot answer' (Derrida, 1978, p. 79). 'It may even be that these questions are not *philosophical*, are not *philosophy's* questions. Nevertheless, these should be the only questions today capable of founding the community, within the world, of those who are still called philosophers.' Henceforth,

> all those boundaries that form the running border of what used to be called a text, of what we once thought this word could identify, i.e., the supposed end and beginning of a work, the unity of a corpus, the title, the margins, the signatures, the referential realm outside the frame, and so forth. What has happened, if it has happened, is a sort of overrun that spoils all these boundaries and divisions and forces us to extend the accredited concept, the dominant notion of a 'text,' of what I still call a 'text,' for strategic reasons, in part – a 'text' that is henceforth no longer a finished corpus of writing, some content enclosed in a book or its margins, but a differential network, a fabric of traces referring endlessly to something other than itself, to other differential traces. Thus the text overruns all the limits assigned to it so far (not submerging or drowning them in an undifferentiated homogeneity, but rather making them more complex, dividing and multiplying strokes and lines) – all the limits, everything that was set up in opposition to writing (speech, life, the world, the real, history, and what not, every field of reference – to body or mind, conscious or unconscious, politics, economics, and so forth). Whatever the (demonstrated) necessity of such an overrun, such a *débordement*, it still will have come as a shock, producing endless efforts to dam up, resist, rebuild the old partitions, to blame what could no longer be thought without confusion, to blame difference *as* wrongful confusion! All this has taken place in nonreading.
>
> (Derrida, 1991, pp. 256–7)

On the basis of this ex-tension of the text – in which reading is lubricated to the point of losing its t(h)read – it should be easy to understand why deconstructive writing perpetually overruns and overspills its allotted space, and always engulfs every conceivable strategy and type of partition. This is why deconstructive writing neither re-presents, nor shows: it flows, circulates, spreads and luxates. In other words, it knows nothing of fixed forms and

Being. To the contrary, it lives through Flux, Multiplicity, Dissemination, Iridescence and Becoming. Instead of thinking in terms of deconstructive writing as writing *about* something (i.e. at a distance from its subject), think of deconstructive writing as circulating *through* something (i.e. as immersed within its subject matter). Consequently, deconstructive 'writing does not have a place for language over here, and a world over there to which it refers' (Bennington, 1989a, p. 84). Or again: 'There is no essential difference between language and the world, the one as subject, the other as object. There are traces. . . . Think of Deconstruction as extending the world paradigm if you like. It makes no difference, so long as you don't think of the world set up out there over against' (Bennington, 1989a, p. 84). Moreover, and as we have already seen, the problem with deconstructive writing is not so much one of crossing or transgressing limits. Rather, the problem is one of locating and approaching limits. Find a text to read if you can. It is for all of these reasons that one can say that (deconstructive) writing 'proliferates *outside* to the point of no longer being *understood*. It is no longer *a* tongue', *a* language, *a* text (Derrida, 1982, p. xv). It risks making sense.

With this motif of proliferation and overrun, deconstructive writing can be seen to be both affirmative and productive: 'Deconstruction is *not* negative, is not nihilistic' (Derrida, 1989b, p. 74). And one might venture the claim that '*in a certain sense* deconstruction is prior to construction' (Bennington, 1989b, p. 86), where this coming before is interminable. It acts as a prefatory disarrangement of the text which will therefore never have been written or read. In short, prefatory deconstruction dismantles and erodes the ground of reading: no ground, no construction; no gravity, no reading! Rorty (1982, 1989, 1991) and Habermas (1987) miss all of this because they begin with an (abysmal) *a priori* separation of world and text. 'Consider Derrida as trying . . . to create a new thing for writing to be about,' suggests Rorty (1982, p. 95), 'not the world, but texts.' It is, of course, only a short step from this confusion to the claim that: 'No constructors, no deconstructors. No norms, no perversions. Derrida (like Heidegger) would have no writing to do unless there were a "metaphysics of presence" to overcome. Without the fun of stamping out parasites, on the other hand, no Kantian would continue building' (Rorty, 1982, p. 108: see Bernstein, 1988). Basically, pragmatism wins out through a bizarre case of misrecognition: everybody needs everybody else.

'My criticisms of Heidegger . . . and of Derrida,' continues Rorty (1991, p. 3), 'centre around their failure to take a relaxed, naturalistic, Darwinian view of language.' Indeed, 'The superiority of later to earlier Derrida seems to me precisely that he stops relying on word magic and relies instead on a way of writing – on creating a style rather than on inventing neologisms' (Rorty, 1989, p. 124, n. 6). On Rorty's account, Habermas's criticism of Derrida is vindicated: 'An aesthetic contextualization blinds him to the fact that everyday communicative practice makes learning processes possible. . . . He permits the capacity to solve problems to disappear behind the world-creating capacity of language' (Habermas, 1987, p. 205). The later 'Derrida privatizes his philosophical thinking, and thereby breaks down the tension between ironism and theorizing. He simply drops theory . . . in favor of fantasizing' (Rorty, 1989, p. 125). Both Habermas and Rorty have failed to grasp the significance of Derrida's insistence on the *materiality* of the text (Derrida, 1981a, 1981b). They have not unfolded the full consequences of their truly abysmal characterisation of 'language,' 'writing,' 'speech' and 'communication.' Indeed, both Habermas and Rorty have failed 'to take back from idealism the "active side" of practical knowledge which the materialist tradition, notably with the theory of "reflection", had abandoned to it' (Bourdieu, 1990, p. 13). Contrary to popular opinion, Derrida's work has been continually directed against the establishment of what he refers to as a new 'theology of the Text' (Derrida, 1981a, p. 258) – and especially the establishment of a negative theology, or what Rorty (1991, p. 116) refers to as 'the worship of a Dark God, the celebration of perpetual absence' (Merquior, 1986; Dews, 1987; Rosen, 1987). 'The materialist insistence,' wrote Derrida (1981b, p. 66), 'can function as means of having the necessary generalization of the concept of text, its extension with no simple exterior limit . . . not wind up, then, as the definition of a new self-interiority, a new "idealism," if you will, of the text.' It is in this sense that the world *is* text.

Deconstructive writing defies the Spirit of Gravity, the Spirit of Reading. It defies the rule, the force and the violence of Necessity. 'I abandon this reading to you' writes Derrida (1988, p. 25). Yet this reading, so cast, does not fall. It does not break its back under the force of the Spirit of Gravity: it floats, tortuously. The abandoned reading is not simply a reading. Between anticipation and frustration, deconstructive writing remains in step, between things. 'One of the best pedagogical approaches to' Derrida's work, notes Bennington (in Derrida, 1989b, p. 78), is 'to try and think less in terms of the "this and that" than in terms of the tension "between."' Being in step, with an interminable preface to a borderless text, between two covers and for the duration of this written distension: '*Between* things does not designate a localizable relation going from the one to the other and reciprocally, but a perpendicular direction, a transversal movement carrying away the one *and* the other, a stream without beginning or end, gnawing away at its two banks and picking up speed in the middle' (Deleuze and Guattari, 1983, p. 58). Or again: deconstructive writing is forever 'Gnawing away at the border' (Derrida, 1982, p. xxiii). Forget reading, then, not because the reading is finished or perfected, and still less because the text has been exhausted ('No danger. We are very far from that; this right here, I repeat, is barely preliminary' Derrida, 1986b, p. 204), but because all of this will have taken place in nonreading. The *is* of the written phrase 'the world is text' perpetually uproots, hollows and displaces that which would be the subject of a reading: becoming other, colloidal and fractal ('Deconstruction is first and foremost a suspicion directed against just that kind of thinking – "what is . . . ?"' Derrida, 1989b, p. 73). After Nietzsche's Zarathustra, Tympani: 'To philosophize with a hammer' (Derrida, 1982, p. xii). Forget reading, then, because we will never have found the limit of a text. Henceforth, there is no(thing) inside or outside the text: *hollow surface, signsponge* ('All the examples are thus cut out and cut across each other. Look at the holes, if you can' Derrida, 1986b, p. 210. Indeed, 'It is invisible hands that torment and bend us the worst' Nietzsche, 1969, p. 69). The fate of reading in general, and of humanism and hermeneutics in particular, is always already sealed through an interminable dispersal. 'It suffices, in sum, barely, to wait' (Derrida, 1982, p. xxix): ' "more passive than passivity," hear it as the most provocative thought today' (Derrida, 1991, p. 424). Truly, the world abandons its landscapes to you (Lyotard, 1989).

Deconstructive writing defies the Spirit of Gravity, the Spirit of Reading. It defies the curvature of a simple, impatient, thrifty, miserly, serious and restricted stride which sees no problem in crossing its limit, its abyss. 'One day you will cry: "Everything is false!"' (Nietzsche, 1969, p. 89), including (and especially) these burdensome and insatiable spirits!

Preface to a short treatise on the pathos of the would-be reader who is forever doomed to wander across inhospitable terrain, without shelter or respite, leaving nothing but footprints on indifferent and asignifying ground: without OME, without limit, without face. The world is text: indifferent, hollow surface, signsponge. I abandon this landscape to you.

PLOUGH TO INFINITY

> And I would say that deconstruction loses nothing from admitting that it is impossible; also that those who would rush to delight in that admission lose nothing from having to wait. For a deconstructive operation *possibility* would rather be the danger, the danger of becoming an available set of rule-governed procedures, methods, accessible approaches. The interest of deconstruction, of such force and desire as it may have, is a certain experience of the impossible.
>
> (Derrida, 1991, p. 209)

> Deconstruction is life to me.
>
> (Derrida, 1989c, p. 223)

In a certain sense, one might say that there are (at least) two genres of deconstruction, one methodological, the other de-limited. Each of these genres can be said to depart from the writings of Jacques Derrida, but neither of which can be said to do justice to his very specific departure. The first genre of deconstruction locates and deploys a philosophical 'method' (particularly within literary criticism) which proceeds through a *reversal* and *reinscription* of the hierarchical oppositions which structure and regulate texts. Derrida (1981b, p. 71) succinctly formalised this method in the slogan: 'extraction, graft, extension.' On the basis of this slogan there has been an attempt 'to pursue . . . a "general economy," a kind of *general strategy of deconstruction*' that would 'avoid simply *neutralizing* the binary oppositions of metaphysics and simply *residing* within the closed field of these oppositions, thereby confirming it' (Derrida, 1981b, p. 41). Moreover, this strategy has sought to 'proceed using a double gesture, according to a unity that is both systematic and in and of itself divided, a double writing, that is, a writing that is in and of itself multiple.' Schematically,

> we proceed: (1) to the extraction of a reduced predicative trait that is held in reserve, limited in a given conceptual structure (limited for motivations and relations of force to be analyzed), *named X*; (2) to the de-limitation, the grafting and regulated extension of the extracted predicate, the name X being maintained as a kind of *lever of intervention*, in order to maintain a grasp on the previous organization, which is to be transformed effectively. Therefore, extraction, graft, extension: you know that this is what I call, according to the process I have just described, *writing*. (Derrida, 1981b, p. 71)

Deconstructive writing attempts to leave a disarranging, material trace in the text by 'lifting' all of those things which have been forgotten, dissimulated, masked, repressed, denied, excluded, buried, hidden and left in reserve, demonstrating that they are essentially (and not marginally) related to the structuration (or possibility) of the text, and that this extended sense must be affirmed. This is the sense in which deconstruction moves from a 'restricted' to a 'general' economy (Harvey, 1986). It seeks to keep nothing in reserve (Hence the almost absolute proximity of dialectics and deconstruction: nothing will be left out – nothing will be forgotten: Derrida, 1978; Lyotard, 1989, 1990). Deconstruction, then, 'is not *neutral*. It *intervenes*' (Derrida, 1981b, p. 93), and one must always 'underline the necessity of reinscription rather than denial' (Derrida, 1981b, p. 94). In this sense, the two gestures of deconstruction are dissymmetrical: whilst reversal is a move which remains entirely *within* the restricted economy; reinscription attempts a breakthrough into the *general* economy without reserve (This is why Gasché, 1986, argues that the dissymmetry of deconstruction ruptures all systems organised on the basis of reflection and representation). But it is important to add that the two gestures of deconstruction take place simultaneously. (And on the basis of dissymmetry and simultaneity, Derrida often deploys the pictogram 'X' to articulate the chiastic intervention of deconstruction). This is why deconstruction is less a way of reading a text, and more a way of writing (on) a (borderless) text (Strictly speaking, the weaving of the borderless text is never finished: find a text to read if you can). In short, deconstructive writing always leaves a material trace or furrow in the text. It *ploughs* through texts with multiple blades (X), turning and overturning relentlessly. 'Hence the heterogeneity of texts' (Derrida, 1979, p. 95). Hence also the fact that 'One never writes in one's own or in a foreign language' (Derrida, 1991, p. 268).

This method of continually turning and overturning is said to provide the grounds (or more precisely, the 'conditions of *im*possibility' as Norris, 1990, p. 200 puts it) for rigorous – therefore very complex (i.e. valorised as good, not bad) – 'interpretation.' For example, Norris argues that deconstruction can be characterised as

> the dismantling of conceptual oppositions, the taking apart of hierarchical systems of thought which can then be *reinscribed* within a different

> order of textual signification. Or again: deconstruction is the vigilant seeking-out of those 'aporias', blindspots or moments of self-contradiction where a text involuntarily betrays the tension between rhetoric and logic, between what it manifestly *means to say* and what it is nonetheless *constrained to mean*. To 'deconstruct' a piece of writing is therefore to operate a kind of strategic reversal, seizing on precisely those unregarded details (casual metaphors, footnotes, incidental turns of argument) which are always, and necessarily, passed over by interpreters of a more orthodox persuasion. For it is here, in the margins of the text, that deconstruction discovers those unsettling forces at work.
>
> (Norris, 1987, p. 19)

Eagleton (1986, p. 80) makes much the same point, suggesting that 'To "deconstruct", then, is to reinscribe and resituate meanings, events and objects within broader movements and structures; it is, so to speak, to reverse the imposing tapestry in order to expose in all its unglamorously dishevelled tangle the threads constituting the well-heeled image it presents to the world.' Similarly, Lodge (1989, p. 88) describes deconstruction as 'not so much a method, more a frame of mind – one that has tirelessly questioned the nature and possibility of meaning through *analysis* of and commentary upon texts – originally philosophical texts, then literary texts.' But as we saw earlier with Rorty, Lodge (1989, p. 90) believes that this version of 'Deconstruction is on the wane. Derrida's own late work has become increasingly whimsical, fictive and difficult to methodise.'

To signal my departure from this first genre of deconstruction, I will make three very brief points. First, the text (notwithstanding the affirmation of its irreducible différance from itself) fails to take account of the asignifying; instead, it contents itself with a pluralisation (the evocation of 'heterogeneity,' 'difference,' 'otherness,' 'alterity,' etc: all of those words which Derrida, 1982, pp. xiii–xiv, said could never function as 'the conceptual headings under which philosophy's border could be overflowed'). In other words, this gesture still assumes that the text speaks (nonsense or babel). In place of a single, coherent and unitary meaning (univocality) we are offered multiple, dissonant and self-contradictory meanings (equivocality). Second, everything still takes place within reading; it is still a question of interpretation (even if that interpretation has become interminable through the departure of a transcendental signified and the loss of determinate boundaries to the text). It is still what Rosen (1987, p. 9) calls 'the decay of theory into interpretation' (a sort of hermeneutic or humanist version of entropy). Last, this gesture contains absolutely nothing which could disrupt Hegelianism. To the contrary, both the motif of reversal and reinscription, and the slogan 'extraction, graft, extension,' could, if taken literally, if acted upon, if formalised into a methodology, just as well be a definition of the work of Hegelianism itself (*Aufhebung*: position, negation, negation of the negation). Indeed, Derrida is so attuned and sensitised to the 'immense enveloping resources of Hegelianism' (Derrida, 1978, p. 251), which 'regularly change transgressions into "false exits"' (Derrida, 1982, p. 135), that he fills the whole of *Positions* (Derrida, 1981b), the text from where the motif and slogan are 'lifted,' with just such warnings (Deconstructive aphorisms, cautions Derrida, 1989a, p. 69, 'can always once more become dialectical moments, the absolute knowledge held in reserve in a thesis or antithesis'). For example, at the end of *Positions* (by way of a P.S.), Derrida makes it clear that he wants to take the risk of not holding a position: 'one can always redefine, beneath the same word (extraction, graft, extension), the concept of *position*. . . . And if we gave to this exchange, for its (germinal) title, the word *positions*, whose polysemia is marked, moreover, in the letter *s*, the "disseminating" letter *par excellence*. . . . I will add, concerning *positions*: scenes, acts, figures of dissemination' (Derrida, 1981b, p. 96: see Megill, 1985).

As long ago as 1980, Derrida referred to 'deconstruction' as 'a word I have never liked and one whose fortune has disagreeably surprised me' (Derrida, 1983, p. 44). In 1983, Derrida notes how the word 'imposed itself' (Derrida, 1991, p. 270), adding that it 'has interest only within a certain context' (Derrida, 1991, p. 275). 'I do not think . . . it is a *good word*', he continues. 'It is certainly not elegant.' From the very beginning, and now more than ever, the word 'deconstruction' must be withdrawn, ploughed and infolded (although it must never be forgotten). But as we shall see, this withdrawal is not without disruptive and irruptive effects.

Placing this withdrawal and infolding of deconstruction in reserve for a moment, I now want to turn to the second genre of deconstruction which works through the de-limitation of philosophical discourses. Simplifying to the extreme, whilst philosophy can locate (and therefore transgress and sublate) a host of boundaries, limits and margins for itself, it cannot locate (and therefore cannot transgress and sublate) the boundary, limit and margin of itself. And without such an edge, philosophy cannot separate and extricate itself from what it would wish to delimit as non-philosophy (Silverman, 1988, 1989). Hence the fact that philosophy always and necessarily lacks a proper domain of which it would be the sole and rightful proprietor. Or again: philosophy will always fail to delimit a territory (of its own) within which its jurisdiction would be legitimate (uncontestable). Accordingly, in this second genre of deconstruction, where philosophy, strictly speaking, is always already structurally de-limited, one encounters 'the impossibility of a *position* which is not already a *relation*, an ex-position to something (someone?) other' (Kamuf, 1991, p. xv). Moreover, this structurally necessary de-limitation, ex-position and ex-ap-propriation of philosophy gives deconstruction 'a radical incompleteness thesis' (Wood, 1990, p. 48: see in particular, Derrida, 1978, pp. 278–93). There is, if you like, within the texture of philosophy, 'an undecidability over what is inside and what is outside, over thresholds' (Llewelyn, 1986, p. xi). In other words, the de-limited, ex-positioned and ex-ap-propriated texture of (non)philosophy cannot (for essential reasons) yield criteria, limits or margins whose deployment would partition the borderless – and therefore continuously variable (analogical and ana-logical) – text. Concerning this borderless text which would wish (for want of a rule: Lyotard, 1988) to cleave itself into so-many partitions, 'one has to accept the fact that one will never have done with the text's difference from itself.... It is in the space of their difference that our interpretative readings risk losing anchor' (Kamuf, 1991, p. 354). Finally, and insofar as the continuously variable, ana-logical texture of (non)philosophy cannot be partitioned into texts, this second genre of deconstruction attempts a breakthrough into 'writing,' 'matter' and the 'asignifying.' But before we pursue this attempt, it is worth clarifying what is at stake in the refusal to engage in (either univocal or equivocal) reading.

> We could call interpretation a method of reading (and writing) which aims at bringing out the meaning/message of the text more clearly perhaps (or just in a different way) than the original text.... Deconstruction does not do this. Rather than help a text fulfil itself by bringing out its real meaning, it looks at just those points at which the text subverts itself, at which the attempt at closure ... breaks down; it attempts to bring out and subvert the conceptual hierarchies that structure the text. What this does destroy is a *facade of unity*, a certain presumption about what a text is, or should be helped to be. But equally it could be said to *bring to life* in the text the forces active in its construction. The text ceases to be a thing containing a meaning, but becomes (again) a struggle between order and chaos, a desperate attempt to exchange its own materiality for a transparency.
>
> (Wood, 1990, p. 58)

Both of the preceding genres of deconstruction clearly engender very complex ramifications, some of which I would like to trace through the withdrawal, infolding and ploughing of 'deconstruction' into the asignifying matter of the 'borderless text.'

In 'Deconstruction is not what you think', Bennington (1989a, p. 84) claims that 'Deconstruction is not a theory or a project. It does not prescribe a practice more or less faithful to it, nor project an image of a desirable state to be brought about.' As Culler (1988, p. 140) puts it: 'if there is a problem about application it is not because deconstruction is too pure to be applied; it is because it is always already applied.' Deconstruction, then, 'is not, strictly speaking, a theory' (Bennington, 1989b, p. 86). Therefore, 'it cannot be taken to prescribe a practice.' With Derrida, deconstructive writing always seeks to abolish the feigned (i.e. forced) distance between what the writing purports to be about and what the writing is actually doing. Indeed, the generalised and borderless text relentlessly gnaws away at all of the frames, limits, creases, folds, margins, barriers and partitions which have been drawn upon in order to hold work and commentary apart. This is why 'it suffices, in sum, barely, to wait' (Derrida, 1982, p. xxix). In other words, deconstructive writing is performative rather than representational; immersed in things rather than reflecting upon them (Deleuze and Guattari, 1988). Indeed, 'One of the distinctive features of Jacques Derrida's writing', suggests Culler (1988, p. 140), 'has been its refusal to develop a general and consistent theoretical metalanguage.' Deconstructive writing flows, circulates and disperses through the hollowed surfaces of innumerable (borderless) texts: 'Expelled, doomed to exodus. Thus their hatred of geophilosophy' writes Lyotard (1990, p. 93); 'mischievous lubricant' writes Boyne (1990, p. 91); 'To write – dissemination' writes Derrida (1981b, p. 86). In this sense, deconstructive writing is assessed neither according to its coherence, nor according to its correspondence. Rather, it is assessed according to its viscosity. On this account, deconstructive writing could be allied to Lyotard's (1984, p. xxiv) definition of the '*postmodern* as incredulity towards metanarratives' (which should be immediately detached from any periodisation or epochalisation: Doel, 1992), with the proviso that it is linked to a dissemination of presence: '*Post modern* would have to be understood according to the paradox of the future (*post*) anterior (*modo*)' (Lyotard, 1984, p. 81). On this basis, it may be possible to enact, with Derrida, a palaeonomy of the word deconstruction. 'I use the word "palaeonomy" to explain the way we should use an old word,' writes Derrida (1989c, p. 224), 'not simply to give up the word, but to analyze what in the old word has been buried or hidden or forgotten. And what has been hidden or forgotten may be totally heterogeneous to what has been kept.'

Where to begin? Certainly not with the question: 'What is deconstruction . . . ?' (Unless, of course, one understood 'is' to signify *Is it happening?* The *it* indicating an empty place to be occupied by a referent. Which is to say, unless the question itself is always already disseminated and cast out: expelled, doomed to exodus). Without wishing to commence a detailed exploration of this particular terrain, suffice it to say that deconstructive writing resists the facile and futile search for an essence, origin or position (from which one could launch the question, 'What is . . . ?' for example) which is not always already subject to de-limitation, ex-position, ex-ap-propriation, dissemination, etc. As Derrida (1991, p. 275) put it: 'All sentences of the type "deconstruction is X" or "deconstruction is not X" *a priori* miss the point, which is to say that they are at least false.'

> All the same, and in spite of appearances, deconstruction is neither an *analysis* nor a *critique*. . . . I would say the same about *method*. Deconstruction is not a method and cannot be transformed into one. Especially if the technical and procedural significations of the word are stressed. It is true that in certain circles . . . the technical and methodological 'metaphor' that seems necessarily attached to the very word 'deconstruction' has been able to seduce or lead astray.
>
> (Derrida, 1991, p. 273)

At this point we could attempt to trace and retrace the steps of Derrida (jumping over many, and we would not be forgiven for that) through a burgeoning stream of precautions, warnings, nuances, clarifications, ramifications, transliterations, convolutions, etc. And perhaps one would encounter, or at least glean, a sense of the borderless condition (and de-position) of the general text (which would not be a text as such, not merely because it would be a polylogue of at least n + 1 voices, but also because it would not be truly made up of signs, still less of signifiers, etc.), with its perpetual overflow, slippage, spillage, dissemination, disarrangement, ex-position, ex-ap-propriation, ex-tension, seriasure, distension, différance (in short, and criss-crossing all of this with at least n + 1 ploughs [X], the eternal recurrence of: extraction, graft, extension, etc.), as the interminable flow and luxation of (deconstructive and deconstructing) writing de-limits and de-capitates numerous hollow surfaces, each with its own specificity and heterogeneous texture, which may prove to be enthralling, perhaps even a little edifying. But once again, all of these (deconstructive and palaeonomic) gestures would *a priori* miss the point. One would be 'left holding (jealously) the shell from which the desired thing has retracted' as Kamuf (1991, p. xxxiv) puts it: the haunting of a barren place by a ghost writer, the haunting of difference by a little 'a.' Moreover, all of these gestures would have to be crossed and re-crossed by all manner of precautions, warnings, etc. to the effect that 'There will be no unique name, even if it were the name of Being. And we must think this without nostalgia' (Derrida, 1982, p. 27). Consequently, it should be apparent that none of this would exhaust the necessary work required to keep up with and retrace the movements of the phrase *Is it (but what?) happening?* And in order to affirm this falling short, to prevent any word, or connection, or relay, or trait, or precaution, or warning, etc. from giving off the illusion that the (deconstructive and deconstructing) writing had arrived at, or indicated, or even affirmed, the presence (or even the existence: disseminated, or otherwise) of its (own) (proper) limit, one would need to draw upon all manner of textual strategies, including, and I underline, polysemia, ambiguity, undecidability, parody, irony, comedy, but even then, after all of these (let's say, for the sake of conversational expediency and to keep the conversation going: now in a relaxed, naturalistic, Darwinian tone) 'delaying tactics,' one would still be forced to ask (or admit, or affirm, or delay, or warn, etc.), concerning this on-going text (of which this sentence would be a good example), 'should I now write it several times, loading the text with quotation marks, with quotation marks within quotation marks, with italics, with square brackets, with pictographed gestures, even if I were to multiply the refinements of punctuation in all the codes, I wager that at the end the initial residue would return' (Derrida, 1988, p. 19).

> To be very schematic I would say that the difficulty of *defining* and therefore also of *translating* the word 'deconstruction' stems from the fact that all the predicates, all the defining concepts, all the lexical significations, and even the syntactic articulations, which seem at one moment to lend themselves to this definition or to that translation, are also deconstructed and deconstructible, directly or otherwise, etc. And that goes for the *word*, the very unity of the *word* deconstruction, as for every *word*.
>
> (Derrida, 1991, p. 274)

'What has happened? In sum, nothing has been said. We have not stopped at any word; the chain rests on nothing; none of the concepts satisfies the demand, all are determined by each other and, at the same time, destroy or neutralize each other' (Derrida, 1978, p. 274). In writing about deconstructive writing the residue which will have returned is the word itself. It is the presumed unity of the word which is perpetually disseminated. In chasing the word 'deconstruction,' the word fell apart. Writing is thus a kind of inverted Hegelianism, where concepts, instead of gathering together into a totality, are perpetually falling apart into a multiplicity.

In a certain sense, we should be indifferent to the word 'deconstruction.' Its fate is marginal, no one should be haunted by ghosts or little 'a's. Indeed, and as we have seen, 'the word has interest only within a certain context, where it replaces and lets itself be determined by such other words as "écriture," "trace," "differance," "supplement," "hymen," "pharmakon," "marge," "entame," "parergon," etc. By definition, the list can never be closed, and I have cited only names, which is inadequate and done only for reasons of economy' (Derrida, 1991, p. 275). The word 'deconstruction,' like any word, is infinitely displaceable and substitutable: moveable, therefore, removable. Indeed, the whole thrust of dissemination is to affirm the *fact* that a word is always already lost, faulty, marginal, enfolded and sacrificed. Henceforth, leave Derrida and forget deconstruction. I abandon this corpse to you (End of story: write on).

Lyotard and Derrida, however, refuse to forget. Indeed, they insist on not forgetting (Bennington, 1988; Carroll, 1987; Readings, 1991). This insistence does not arise out of nostalgia. Rather, it arises out of a refusal to be in complicity with an insidious, thrifty, miserly and restricted political economy of thought which works through a violent 'forgetting' or 'silencing.' Simplifying to the extreme, the fate of deconstruction, of the word 'deconstruction,' bears witness to the possibility of disappearing without trace. It bears witness to the possibility of being erased. Moreover, it bears witness to the impossibility of proving one's disappearance. Through writing on, we have come to the brink of erasing the word without trace. The fate of deconstruction, of the word 'deconstruction' – and of all the words to which it is attached – demands 'forms of thinking and writing that do not forget "the fact" of the forgotten and the unpresentable' (Carroll, 1990, p. x). For Lyotard (1990) and Derrida, deconstructive writing seeks to affirm the fact of the forgotten, and the fact that the forgotten will still be forgotten even when something is drawn upon in order to 'represent' it. For Lyotard and Derrida, deconstructive writing is obliged to carry and affirm the burden of this fault.

Deconstructive writing is irreducibly faulty. 'There you are, forewarned: it is the risk or chance of that fault that fascinates or obsesses me at this very moment, and what can happen to a faulty writing, to a faulty letter' (Derrida, 1991, p. 409). Writing, all writing, including this writing, is guilty of silencing everything which has not been phrased and everything which will never have been phrased. And yes, we are obligated to extract, graft and extend the 'silenced' and the 'forgotten' from the margins, the lower strata, the shadows, the folds and the creases. And yes, we are obligated to relentlessly plough through all manner of materials, lifting and reinscribing innumerable voices, gestures, meanings, phrases and positions – everything which has been *held* in reserve by a repressive and restricted economy. Indeed, writing is obligated to the forgotten and the silenced (End of story). But this deconstructive writing would not go far enough in its obligation unless it recognised that the silenced and the forgotten have not only been hidden, masked, buried, dissimulated and reserved by a thrifty, miserly, serious and restricted political economy of thought. They have also been erased, expunged, liquidated, annihilated and exterminated by a desire for totalisation.

> One of the fundamental responsibilities of thought is this debt to the Other, an obligation demands that thought become less and less philosophical and more and more 'written' . . . a writing of (and as) exile, wandering, rootless . . . one that never forgets that there is the forgotten and never stops writing its failure to remember and to fashion itself according to memory.
>
> (Carroll, 1990, p. xxiii)

In short, deconstructive writing must not fulfil the desire for a 'final solution' to the 'problem' of the 'fact' of the 'forgotten' (Lyotard, 1988, 1990). It must not allow itself to become the mechanism through which the debts and obligations to the silenced and the forgotten are liquidated, cancelled or denied. It suffices, in sum, barely, to wait. 'If I had to risk a single definition of deconstruction, one as brief, elliptical, and economical as a password, I would say simply and without overstatement: *plus d'une langue* – more than one language, no more of one language' (Derrida, 1986a, pp. 14–15). For Lyotard and Derrida, all we have is a writing against writing which struggles to do justice to that which is silenced and forgotten in (the duration of) writing. Writing must wait for the (non)arrival of the forgotten, the silenced and the other. Therefore, it *does not* suffice, in sum, barely, to wait. Deconstructive writing must become active and affirmative; it is obligated to make space and volume available for the other. This is why deconstructive writing is articulated through the distension of a mute and asignifying hollow surface. Henceforth, the borderless text should be understood as a Stateless text: made to be available to the rootless, the expelled, the annihilated, the silenced and the forgotten – anyone, in other words, marked with the anonymous *Name X*. And as Derrida has demonstrated, all of our words are marked with and obligated by the anonymous *Name X*. 'Let us wage war on totality; let us be witnesses to the unpresentable; let us activate the differences and save the honour of the name' (Lyotard, 1984, p. 82).

The other can only arrive, then, 'by means of a series of words that are all faulty' confides Derrida (1991, p. 424). And insofar as these words are all at fault, all implicated and marked by the innumerable silences of expulsion, forgetting and annihilation, it would be tempting to erase, silence and forget them in turn. 'This is why it is inexact to say that I have erased these words. . . . I should have let them be drawn into a *series* (a stringed sequence of enlaced *erasures*), an interrupted *series*, a *series* of interlaced interruptions, a series of *hiatuses*. . . . that I shall henceforth call . . . the *seriasure*' (Derrida, 1991, p. 424). This is why deconstruction is wholly affirmative. But will this *seriasure*, this writing against writing, this folding and refolding, this ploughing and dredging, this war against totality, against Hegelianism, against the idling reader, suffice for the other to come and be heard?

> I no longer know if you are saying what his work says. Perhaps that comes back to the same. I no longer know if you are saying the contrary, or if you have already written something wholly other. I no longer hear your voice, I have difficulty distinguishing it from mine, from any other, your fault suddenly becomes illegible to me. Interrupt me.
>
> (Derrida, 1991, p. 438)

I am obligated by something I have forgotten to cover. Henceforth, I abandon this borderless text to an anonymous *Name X*, a would-be reader of contemporary rural change.

Find a text
To read
If you can.

REFERENCES

Barnes, T. J. and Duncan, S. J. (eds.) (1992) *Writing Worlds: Discourse, Text and Metaphor in the Representation of Landscape*, Routledge, London.

Baudrillard, J. (1975) *The Mirror of Production*, Telos, St Louis.

Baudrillard, J. (1981) *For a Critique of the Political Economy of the Sign*, Telos, St Louis.

Baudrillard, J. (1983) *Simulations*, Semiotext(e), New York.

Baudrillard, J. (1986) The year 2,000 will not take place, in E. A. Grosz, T. Threadgold, D. Kelly, A. Cholodenko and E. Colless (eds.) *Future *Fall: Excursions into Post Modernity*, Power Institute of Fine Arts, University of Sydney.

Baudrillard, J. (1987) *Forget Foucault: Forget Baudrillard*, Semiotext(e), New York.

Baudrillard, J. (1988) *Jean Baudrillard: Selected Writings*, Polity Press, Cambridge.

Baudrillard, J. (1990a) *Revenge of the Crystal: Selected Writings on the Modern Object and its Destiny, 1968–1983*, Pluto Press, London.

Baudrillard, J. (1990b) *Cool Memories*, Verso, London.

Baudrillard, J. (1990c) *Seduction*, Macmillan, London.

Bennington, G. (1988) *Lyotard: Writing the Event*, Manchester University Press.

Bennington, G. (1989a) Deconstruction is not what you think, in A. Papadakis, C. Cooke and A. Benjamin (eds.) *Deconstruction: Omnibus Volume*, Academy Editions, London.

Bennington, G. (1989b) Deconstruction and postmodernism, in A. Papadakis, C. Cooke and A. Benjamin (eds.) *Deconstruction: Omnibus Volume*, Academy Editions, London.

Bernstein, R. J. (1988) Metaphysics, critique, and utopia, *Review of Metaphysics*, Vol. 42, pp. 225–73.

Bhaskar, R. (1989) *Reclaiming Reality: A Critical Introduction to Contemporary Philosophy*, Verso, London.
Bourdieu, P. (1990) *In Other Words: Essays Towards a Reflexive Sociology*, Polity Press, Cambridge.
Boyne, R. (1990) *Foucault and Derrida: The Other Side of Reason*, Unwin Hyman, London.
Carroll, D. (1987) *Paraesthetics: Foucault, Lyotard, Derrida*, Methuen, London.
Carroll, D. (1990) Forward: the memory of devastation and the responsibilities of thought: 'And let's not talk about that', in J-F. Lyotard (ed.) *Heidegger and 'The Jews'*, University of Minnesota Press, Minneapolis.
Critchley, S. (1990) Writing the revolution: the politics of truth in Genet's *Prisoner of Love, Radical Philosophy*, Vol. 56, pp. 25–34.
Culler, J. (1988) *Framing the Sign: Criticism and its Institution*, Blackwell, Oxford.
Deleuze, G. and Guattari, F. (1983) *On the Line*, Semiotext(e), New York.
Deleuze, G. and Guattari, F. (1988) *A Thousand Plateaus: Capitalism and Schizophrenia*, Athlone, London.
Derrida, J. (1976) *Of Grammatology*, Johns Hopkins University Press, Baltimore.
Derrida, J. (1978) *Writing and Difference*, University of Chicago Press, London.
Derrida, J. (1979) Living on: borderlines, in H. Bloom, P. de Man, J. Derrida, G. H. Hartman and J. H. Miller (eds.) *Deconstruction and Criticism*, Seabury Press, New York.
Derrida, J. (1981a) *Dissemination*, University of Chicago Press.
Derrida, J. (1981b) *Positions*, University of Chicago Press.
Derrida, J. (1982) *Margins of Philosophy*, Harvester Wheatsheaf, Hemel Hempstead.
Derrida, J. (1983) The time of a thesis: punctuations, in A. Montefiore (ed.) *Philosophy in France Today*, Cambridge University Press.
Derrida, J. (1986a) *Mémoires for Paul de Man*, Columbia University Press, New York.
Derrida, J. (1986b) *Glas*, University of Nebraska Press, Lincoln.
Derrida, J. (1987) *The Post Card: From Socrates to Freud and Beyond*, University of Chicago Press.
Derrida, J. (1988) The truth in painting, in A. Papadakis (ed.) *The New Modernism: Deconstructionist Tendencies in Art*, Academy Group, London.
Derrida, J. (1989a) Fifty-two aphorisms for a foreword, in A. Papadakis, C. Cooke and A. Benjamin (eds.) *Deconstruction: Omnibus Volume*, Academy Editions, London.
Derrida, J. (1989b) Jacques Derrida interview by Christopher Norris: discussion and comments, in A. Papadakis, C. Cooke and A. Benjamin (eds.) *Deconstruction: Omnibus Volume*, Academy Editions, London.
Derrida, J. (1989c) On colleges and philosophy: Jacques Derrida with Geof Bennington, in L. Appignanesi (ed.) *Postmodernism: Documents 4/5*, Free Association Books, London.
Derrida, J. (1991) *The Derrida Reader: Between the Blinds*, Harvester Wheatsheaf, Hemel Hempstead.

Dews, P. (1987) *Logics of Disintegration: Post-Structuralist Thought and the Claims of Critical Theory*, Verso, London.
Doel, M. A. (1992) In stalling deconstruction: striking out the postmodern, *Environment and Planning D: Society and Space*, Vol. 10, no. 2, pp. 163–80.
Eagleton, T. (1986) *Against the Grain: Essays 1975–1985*, Verso, London.
Ellis, J. M. (1989) *Against Deconstruction*, Princeton University Press.
Gane, M. (1991a) *Baudrillard: Critical and Fatal Theory*, Routledge, London.
Gane, M. (1991b) *Baudrillard's Bestiary: Baudrillard and Culture*, Routledge, London.
Gasché, R. (1986) *The Tain of the Mirror: Derrida and the Philosophy of Reflection*, Blackwell, Oxford.
Habermas, J. (1987) *The Philosophical Discourse of Modernity: Twelve Lectures*, Polity Press, Cambridge.
Harvey, I. (1986) *Derrida and the Economy of Différance*, Indiana University Press, Bloomington.
Kamuf, P. (1991) Introduction, in J. Derrida (1991) *A Derrida Reader: Between the Blinds*, Harvester Wheatsheaf, Hemel Hempstead.
Kellner, D. (1989) *Jean Baudrillard: From Marxism to Postmodernism and Beyond*, Polity Press, Cambridge.
Llewelyn, J. (1986) *Derrida on the Threshold of Sense*, Macmillan, London.
Lodge, D. (1989) Deconstruction: a review of the Tate Gallery Symposium, in A. Papadakis, C. Cooke and A. Benjamin (eds.) *Deconstruction: Omnibus Volume*, Academy Editions, London.
Lyotard, J-F. (1984) *The Postmodern Condition: A Report on Knowledge*, University of Manchester Press.
Lyotard, J-F. (1988) *The Differend: Phrases in Dispute*, Manchester University Press.
Lyotard, J-F. (1989) *The Lyotard Reader*, Blackwell, Oxford.
Lyotard, J-F. (1990) *Heidegger and 'the jews'*, University of Minnesota, Minneapolis.
Megill, A. (1985) *Prophets of Extremity: Nietzsche, Heidegger, Foucault, Derrida*, University of California Press, London.
Merquior, J. G. (1986) *From Prague to Paris: A Critique of Structuralist and Post Structuralist Thought*, Verso, London.
Nietzsche, F. (1969) *Thus Spoke Zarathustra*, Penguin, Harmondsworth, Middlesex.
Norris, C. (1987) *Jacques Derrida*, Fontana, London.
Norris, C. (1990) *What's Wrong with Postmodernism: Critical Theory and the Ends of Philosophy*, Harvester Wheatsheaf, Hemel Hempstead.
Readings, B. (1991) *Introducing Lyotard: Art and Politics*, Routledge, London.
Rorty, R. (1982) *The Consequences of Pragmatism*, Harvester Press, Brighton.
Rorty, R. (1989) *Contingency, Irony, Solidarity*, Cambridge University Press.
Rorty, R. (1991) *Essays on Heidegger and Others: Philosophical Papers*, Volume 2, Cambridge University Press.
Rosen, S. (1987) *Hermeneutics as Politics*, Oxford University Press.
Silverman, H. J. (ed.) (1988) *Philosophy and Non-Philosophy since Merleau-Ponty*, Routledge, London.

Silverman, H. J. (ed.) (1989) *Derrida and Deconstruction*, Routledge, London.
Tallis, R. (1988) *Not Saussure: A Critique of Post-Saussurean Literary Theory*, Macmillan, London.
Wood, D. (1987) Following Derrida, in J. Sallis (ed.) *Deconstruction and Philosophy: the Texts of Jacques Derrida*, University of Chicago Press, London.
Wood, D. (1990) *Philosophy at the Limit*, Unwin Hyman, London.

Chapter 4

(En)culturing Political Economy: A Life in the Day of a 'Rural Geographer'

Paul Cloke

I

> Critics need to let go of their distanced and false stance of objectivity and . . . expose their own point of view – the tangle of background, influences, political perspectives, training, situations, that helped form and inform their interpretations.
>
> (Christian, 1989, p. 67)

Over recent years this challenge of exposing and unravelling the personal point of view, and therefore the person, has become more pressing for me in my work as a 'geographer' with longstanding research interests in rurality, rural change and rural governance. Along with so many others, I have become convinced that my writings and readings about the rural have reflected a particular tangle of influence, politics, morality, training, faith and situation which Christian speaks about. Moreover the tangle has changed in nature, sometimes subtly, sometimes radically, during the nineteen years I have been involved with 'rural' research and so, thereby, have my interpretations of the rural also changed. I have no doubt, then, that the process of self-reflection is important in understanding authorship and readership, but also in understanding the changing trajectory and cadence of interpreting 'other' subjects.

It is one thing to acknowledge this (and to recommend it to students for their 'benefit'), and quite another to write specifically in a self-reflective mode. This book, however, presents an opportunity for each of the authors involved to offer an account of the important differences in their preconceptions about, and attitudes towards rural change. For myself, it has led to considerable and critical reflection on just what influences, experiences, theories, socialities and spatialities have over the years fired my geographical imagination and interpretation of people and places that have been labelled as rural. Such reflection has been encouraged by the recent espousal of ethnography as not only a legitimate but also a sensitive method of practising (in my case) human geography.[1] It has also been specifically influenced by a reading of particular literatures on ethnography and self-reflection. It bothered me greatly for a while that academic autobiography appears to be targeted as being mere self-aggrandisement, with the received emphasis being on the 'I' and on the apparent need of the 'I' to legitimise personal actions and texts. Such criticisms can (and indeed sometimes seem designed to) turn researchers away from ethnography, and at the very least they induce a heightened sense of vulnerability in those seeking to acknowledge the self as integral to the processes of constructing and interpreting texts. So is it possible to be self-reflective without being egotistic? I have greatly benefited in this context from reading Elspeth Probyn's (1993) book *Sexing the Self* in which in the context of discussing gendered positions in cultural studies she beautifully summarises previous suggestions that the ethnographic project is about a *production* of the real rather than about how to discover it, and that ethnography thereby concerns the practice of *writing* culture rather than revealing it (see also Barnes and Duncan, 1992). In this way, interpretation can be seen as story-telling, as Steedman (1986) suggests: 'Once a story is told, it ceases to be a story: it becomes a piece of history, an interpretative device' (p. 143).

In this chapter, then, I offer a brief autobiographical account of the changing nature of interpreting rurality contained within my own readings of and writings on the 'rural'. What follows might, then, be labelled at least in part as autoethnography, and it is therefore important at the beginning to place this account within the wider debates over the nature and purpose of ethnography. Without retelling what has already been told extensively elsewhere (see, for example, Atkinson, 1990; Layder, 1993; Stanley, 1990) it seems crucial to acknowledge some of the dilemmas in ethnography which arise from tensions between the ethnographer's self and the ethnographer's attempt to describe 'other' subjects. In popular terms at least ethnography is both a research practice and an interpretative practice. As Grossberg (1989) explains: 'Ethnography is, in the first instance a certain kind of practice in the field, although it is not clear what sort of practice. Additionally, ethnography is a writing practice in which the other is inscribed within, and explained by, the power of the ethnographer's language' (p. 23).

As a research practice, especially in anthropology (see Radway, 1988, 1989) but also in other disciplines, ethnography has been used to present a description of the daily practices of people both in their social situations and in the wider context of history and culture. Such descriptions have had to cope with the idea that different individuals will have different, and even competing, experiences and will represent themselves differently. They also have to deal with the problems of interpreting what is a 'true' description and the 'strangeness' or exotic nature of other subject(s). As a writing practice, ethnography can become a purely discursive activity, fully constructed through the ethno-

[1] Here I would wish to acknowledge my thanks to Phil Crang, Mark Goodwin, Joe Painter, and especially Chris Philo – my co-authors of *Practising Human Geography* – and to my friend and colleague Nigel Thrift, who have variously pointed me towards literatures on these themes.

grapher. The work of James Clifford (1986) fits this description, and in his claiming the ground of 'the novel or avant-garde cultural critique' (1986, p. 23) for ethnography, he privileges the self through textual problematics of representation and discourse over the equally problematic idea of interpreting the 'real' lifestyles of other 'real' people.

Some ethnographies have sought to engage with both research practice and writing practice. Clifford Geertz (1988) for example stresses the need for (anthropological) ethnographers to 'be there' in the field in order that what is written is drawn from and informed by real encounters with the other. According to Geertz, then, the project for ethnographers is to convince us that 'What they say is a result of their having penetrated (or, if you prefer, been penetrated by) another form of life, of having, one way or another, truly been there' (p. 4). Thus, the writing, and the being there, are essentially interconnected. On the one hand this ethnography offers scope for a sensitivity of interpretation – allowing people to speak with their own voices and to articulate their own categories of experiential concern –; on the other hand it offers equal scope for exclusion of the other, as is clearly shown in Said's (1979) account of the discursive and economic appropriation of Asiatic orientalism by the western traders, soldiers, missionaries, etc. who were privileged by 'being there'.

Elspeth Probyn's contribution to this literature is to focus on the self-reflective moment in ethnography. She explains that 'the reason I find myself here is to problematize the operations of the ethnographer's self' (p. 62) and she is keenly concerned with key questions about the voyeuristic self which ethnography allows to be verbalised and reflected upon: 'In short, questions about where one can speak from, to whom one speaks, and why one speaks at all seem to be more immediately articulated within ethnography than elsewhere' (Probyn, 1993, p. 61). These are the very questions with which my own self-reflection on nineteen years of reading and writing rural geographies began.

The emerging literatures on autoethnography emphasise a number of tensions which it is worth giving brief discussion to here because these self same tensions surface repeatedly throughout the chapter in terms of my self and my story-telling and interpretation of the rural. First, the remaking of history through interpretative story-telling will be characterised by an essential tension between how the story is read and how it is written (see Bristow, 1991). Both the reading and the writing are mobile positions, and they are interconnected by multifaced interpretative practices. Spivak (1989) has presented a lively discussion of these ideas of reading and writing, and her thoughts implicate 'reading' and 'writing' of texts into our very being:

> We produce historical narratives and historical explanations by transforming the socius upon which our production is *written* into more or less continuous and controllable bits that are *readable*. How these readings emerge and which ones get sanctioned have political implications on every possible level. . . . Writing and reading in such general

> senses mark two different positions in relation to the many-strandedness of being.
>
> (Spivak, 1989, pp. 269–70)

The scope for the egotistical 'I' is very clear here. Writing will involve the decision – explicitly or implicitly – of whether the author wishes to elevate her or his story above others, thereby producing a 'model' for others to follow, or whether the story is to be part of a more pluralist landscape of many different narratives.

Probyn's (1993) discussion of self-reflexivity takes up this question. If self-reflexivity represents a questioning of the subjectivity of the author/researcher in the process of attempting to understand 'other' subjects, then the ontology of self-reflexivity will be crucial in establishing not only how to tell stories and which stories to tell but also who is telling whose stories, for what purpose and from what standpoint. To use her words, 'we need to ask what exactly a self-reflexive self is reflecting upon . . . it needs to be clear where that self is positioned and whether it is a textual or physical entity' (p. 62). She proposes therefore, an ethnography in which there is an essential tension of experience between the ontological and the epistemological:

> Experience can be made to work in two registers: at an ontological level, the concept of experience posits a separate realm of existence – it testifies to the gendered, sexual and racial facticity of being in the social; it can be called an immediate experiential self. At an epistemological level, the self is revealed in its conditions of possibility; here experience is recognised as more obviously discursive and can be used overtly to politicize the ontological. Both of these levels – the experiential self and the politicization of experience – are necessary as the conditions of possibility for alternative speaking positions within cultural theory.
>
> (Probyn, 1993, p. 16)

Probyn's project is to accept that self-reflection will be drawn simultaneously from direct individual experience and from placing such experiences into particular pre-existing categories of experience (or 'essentialist spaces'). She follows Diana Fuss (1989, p. 118) in recognising that 'we need both to theorise essentialist spaces from which to speak and, simultaneously, to deconstruct those spaces to keep them from solidifying' in order to make productive use of the self in processes of interpretative reflection.

Probyn's work, therefore, suggests that it is important to acknowledge both personal positionality, and broader categories of discourse which will reflect these essentialist spaces. Positionality is by now a familiar concept and indeed it is the point at which this chapter began with Barbara Christian's idea of the tangle of background, influences, political perspectives, training and situations which help form and inform our interpretations. There is a real danger here that positionality within self-reflexivity will itself become a 'hidden form of self-indulgence' (Ang, 1989) – a way in which the personal platform for interpretation can be reinforced rather than questioned. Indeed Probyn herself expresses doubts about how one actually goes about *doing* positionality in theory or in practice, but her contribution to resolving this doubt is significant in the suggestion that the self should be a speaking position which entails a 'defamiliarisation of the taken-for-granted' (p. 80). This means a constant epistemological questioning of how one speaks, and an uneasiness when speaking the self because any assurance of ontological importance should be regularly challenged. It also means that speaking the self should be done modestly and 'from an angle that skewers the inflation of the academic ego' (p. 81).

Discourse analyses are also by now familiar elements of the interpretative landscape (see, for example, Barnes and Duncan, 1992). In the rural studies literature, for instance, Halfacree (1993) has discussed the interconnections between academic and lay discourses of rurality. However, the autoethnography required to unravel the interdiscourse between those discourses contained within various texts, and those through which we make sense of our 'selves', our social experiences and our social relations is complex and multifaceted. Fiske (1990) introduces a discussion of such autoethnography by emphasising that these various discourses 'worked not only to circulate meanings but also to constitute "me" as both a social agent in the reproduction and regeneration of those meanings, and also as the social agency through which they circulated' (p. 85). In his ethnographic study of viewer readings of a TV programme, *The Newly Wed Game* (Fiske 1989), he discerned in himself three main discursive practices: from which meanings were drawn:

(i) a professional discourse, which he says 'blurred the distinction between the domestic and the academic',
(ii) a more popular discourse, in which his taste tended to contradict that expected from a white, middle-class male,
(iii) a bundle of semantic discourses (chiefly gender, but also age, class and race) with which he made sense of the multitude of topics which infused his daily life.

Fiske argues that:

> This meaning process is cultural even though I trace it in the realm of the personal because the sense I make of myself is not only reproduced and regenerated from social resources, both discursive and textual, but is then inevitably put into social circulation in myriad ways as I move through my daily histories and encounters. My first investigation, then, was of myself, not as an individual, but as a site and as an instance of reading, as an agent of culture in process – not because the reading I produced was in any way socially representative of, or 'extrapolable' to, others, but because the process by which I produced it was a structured instance of culture in practice.
>
> (Fiske, 1990, p. 86)

The transfer of personal meanings into cultural processes will occur explicitly in the various disseminative texts which researchers, writers and teachers produce as part of their everyday practice. Fiske reminds us of the difficulties in distancing ourselves from these various discourses in order to describe and theorise them: we have to be able to move in and out of domestic environments, to bring different and distancing discourses to bear on our experience, to make that experience both public and private, and to explain it both as a particular cultural practice and as a precise instance. When such meanings are then used to interpret that other than ourselves, 'ethnographic accounts are portraits, the image of the other a palimpsest through which we see ourselves' (p. 81).

It is with a sense of excitement that a recognition of positionality and discursive meaning can provide important testimony to the understanding of geographical imagination, and with a sense of apprehension because of the potential pitfalls outlined above that I now begin my attempt to describe the various influences, experiences and practices which have accompanied my changing interpretations of rurality and rural change over the last nineteen years.

II

I was born in the North London suburb of Enfield into what I thought of for quite a long time afterwards as a very normal white middle-class household. My dad was a local government officer and then a college administrator – employment which he saw as necessitating strict political neutrality, meaning that politics was not only banned from the front window during election times, but was also rarely discussed with my brother and myself. My mum was a 'housewife' and then later worked part-time at a school, but her real joy was in botany. They were (and are) active Christians and so our family life was punctuated by church services, attendance at church-based organisations (such as the Boys Brigade) and the practice of church-based hospitality (particularly to visiting 'missionaries'). It was a happy and uncomplicated home life with lots of sport and (loud) music and regular high-family-profile annual holidays to inexpensive locations in Britain.

I'm sure that these first eighteen years must have had an enormous impact (positively and negatively) on the way that I think, react and interpret, both explicitly and implicitly: gender roles and sexuality were 'orthodox'; politics were quietly conservative; attitudes towards ethnicity and colonialism were latently 'orthodox' for the time, but overlain by a strong Christian concern for those in need; and a fortnight in Pitlochry was exotic. My brother is now a teacher of geography, and we've often wondered how it was that we both ended up in the same subject area. Perhaps it was mum's feel for nature and the natural environment that was transferred across, or perhaps we responded to similar stimuli in the teaching of geography at school. Or was it the obvious interest in maps, wayfinding, camping and hill walking that was succoured by the Duke of Edinburgh award scheme?

Whatever the exact influence I went off to Southampton University to do a Geography degree with no great attachments to radical causes except for a commitment to Christian faith and ethics, but even this was not worked out in terms of attitudes towards the social and economic problems of the day. This was a time in which a number of radical issues floated past, both academically and socio-politically without ever making much of an impact. I remember, for example, a Student Union debate on whether or not to send financial aid to the Black Panthers, but this was an issue which grabbed my imagination less than that of whether or not I could 'get away with', socially, walking out of an E.L.O. gig because from our position in front of the speaker-stack it was just plain too loud. These were times when self was subjugated at least in part both to the social mores of the day and to the teacher–student relationship in terms of academic work.

I remember relatively little in detail about my geography course. I suppose in some other places the period 1971–4 would have seen the introduction of a radical edge to the subject, but if it was there in Southampton I wasn't being receptive to it. What did seem to be an emphasis in the Department which could be grasped was the idea that planning was both a legitimate target for social scientific endeavour and a process which was clearly capable of improving the situation in 'problem' areas. Firstly in urban and regional contexts, then later in the rural environment, we were introduced persuasively to the notion of progress through planning to a better, more efficient world. Matters of equity were probably also included in this equation, but I do not recall these being a major personal influence. Thus modernity became normalised in my geographical imagination at this early age, linking almost surreptitiously with the political interventionism of the time.

It was in a third-year course at Southampton that I was introduced to, and picked up a fascination for, rural areas and the changes taking place therein. Talking with others who have pursued academic careers it seems that the influence of a particular teacher is often strategic in establishing a field of interest. In my case it was Brian Woodruffe, who through the personable leading of a fieldcourse to Val d'Herens in Switzerland, and a rural planning course which was taught as if he knew exactly what it was like to live in amongst rural change (which indeed he did), who sparked my interest in the rural. I think that at this stage I hadn't interrogated the idea of rurality at all. It was the given 'out there' area beyond the city, to which excursions could be made to see how people lived and how things changed; it was perhaps still the exotic to the ordinariness of a suburban upbringing in North London.

Part of Brian Woodruffe's course dealt with rural settlement planning strategies, and here the notion of key settlements was mentioned. It seemed as though local authorities, as a response to the economic pressures of the time and as a way of 'tightening up' on planning in the countryside, were selecting particular villages for housing growth and service investment so as to build up pseudo-central places with which to service and maintain the settlement pattern. Attracted to the idea that basic theories were being put into practice via the planning system in rural areas I proposed this as a potential Ph.D. topic and sought advice about where would be an appropriate research school to carry out this kind of work. At this time the figureheads of rural planning thought and research were Gerald Wibberley and Robin Best at Wye College, and so I applied to and ended up at Wye to work with Robin Best on the 'key settlements' theme.

Wye was the first village I ever lived in, though even this exposure to the rural 'idyll' was delayed by a year living in Canterbury. Even Wye, I soon got to realise, was not 'typical' rural living – with a railway station at the end of the street which would get you to London in an hour and twenty minutes, and with the dominance of a College with its professional, academic and student discourses very much to the fore. The Countryside Planning Unit in which I was based was an outlier to the specialism in agricultural economics, which in turn was surrounded by a sea of other agricultural sciences. Here as a graduate student my immediate social network, and informal academic debating society, was made up of an economist researching the externalities of 'farm smells', environmentalists specialising in ornithology and eco-habitats, and a range of plant and animal scientists who did things in greenhouses and on the farm. Reading what I have just written, it all sounds rather disparaging, but the opposite is true in that my own prejudices towards particular forms of social science and ignorance of scientific practice and research strategy were stretched (and often ridiculed!) in this environment.

This was not a locus in which emerging radical perspectives in social science were the subject of much, if any discussion. In retrospect I now realise that each in their different ways Gerald Wibberley, Robin Best, and especially Michael Redclift could have passed on so much about political pragmatics and theories about which I knew little, but which would have prompted very different story-telling interpretations in my thesis. Whether for want of research training, or that kind of inquiring personality, however, I just didn't grasp the opportunity to ask the right questions of the right people, and so my research was traditionally and narrowly focused rather than identifying with the conceptual and political issues of the day.

Responding to the cultural prompting that a social science thesis should display statistical competence, I spent a considerable amount of time on two statistical projects. The first was to construct an 'index of rurality' using principal components analysis. Given my view now that this work is an inappropriate way of addressing the idea of what and where is rural, I have often asked the question of why I did this indexing. My empirical work on evaluating key settlement policies was focusing on parts of Devon (which I constructed as a 'remoter' rural area) and Warwickshire (a 'pressured' rural area) and so although I persuaded myself otherwise, the index was not necessary as a mechanism for selecting case study areas. Apart from the 'prevailing social science culture' which legitimised and maybe even necessitated this sort of thing, I can only suggest that I was expressing a rather naive interest in the *question* of what 'rural' was/is in the only way that at the time I had the academic and cultural competence so to do. I think that I knew at the time that by selecting a number of variables to represent, collectively, the rural I was pre-determining the outcome, but the interest was in the emerging geographies of that pre-determination. Ironically, this area of story-telling has been used quite widely by other rural story-tellers – a factor I'm ashamed to say that partially led to a replication of the work ten years later using fresh census material. Incidentally, the second statistical project was the use of factorial surfaces to describe varying levels of 'opportunity' in the case study areas. This story, however, has been safely restricted to the gaze of my external examiner, my supervisor and myself.

During my three years at Wye College I read widely in the planning literatures, became a little more confident in researching and writing, and got to know parts of (but not necessarily the people of) Devon and Warwickshire very well. I had strengthened my attachment to planning discourses and ethics, at least partly because of Viv's work as a planner for Kent County Council. I remained, however, firmly set in the positivistic and planning traditions of human geography research. Neither a rural lifestyle, nor introduction to practical political issues, nor immersion in new literatures prompted any great self-expression. To a large extent I was having to/trying to work things out for myself, and looking back it seems a slow process for the self to emerge from the orthodox 'given' in which it was submerged.

III

It would be all too easy to describe my changing interpretations of the rural as a modernistic journey from darkness to light, ignorance to knowledge, repression to freedom, not-thought-out to fully-thought-out. Such a travelogue is certainly not the intention here, although the metaphor of travel does seem to be appropriate. As Grossberg has said:

> The travel metaphor seems quite appropriate to ethnography. To put it simply, ethnography is always about traversing the difference between the familiar and the strange. The ethnographer leaves her home (the familiar) and travels to the other home (the strange) and then returns home to make sense of it in her writing.
>
> (Grossberg, 1989, p. 23)

In this way I feel that particular travels to 'strange' literatures and contact with 'strange' ideas have been symptomatic of the changing ways in which I have returned home and tried to make sense of the strange in my writing. Thus the journey is not one of progress towards ultimate knowledge or interpretation, but one in which a familiarity with that which was previously strange influences a different form of interpretative story-telling of the subject(s) concerned. So it has been with my continuing engagement with the rural.

After leaving Wye College I started work as lecturer in geography at St David's University College, Lampeter, in west Wales – an institution which David Thomas (the founding Professor of Geography) had nicknamed 'a Wye in the hills'. Here, though, was a very different form of rural lifestyle than that in Kent. As Viv and I looked for a house to buy, a local estate agent proudly outlined the advantages of Lampeter, chief amongst which was that 'it is a nice area because there are no blacks here'. I have to admit that we were *astounded* by the comment, and discussed it with people for weeks afterwards. Of course there had been quite a few black people in Wye, but these had been 'overseas' students at the college rather than accepted full-time members of the community. Were all rural areas prone to this kind of racism then? Had our lives really been that sheltered? The answer to the latter question is probably 'yes'.

There were other distinctive facets of a lifestyle in rural west Wales which also opened up experientially new traverses into the rural unfamiliar. It is a Welsh-speaking area and local cultural organisation and circulation are through this medium. Our experience was that local people were enormously hospitable and friendly towards us, but that even though Viv learned to speak the language neither of us deep down could call Lampeter 'home'. This sense of not belonging found fervent political expression in the more radical nationalist slogans – 'come home to a real fire, buy a second home in Wales', and 'if you want to solve all the problems in Wales, move the English out'. This muddle of interpersonal warmth amongst some political hostility was experienced differently by different people, but was highlighted for us when the National Eisteddfod came to Lampeter, and our 'home' was swamped by the vibrant celebrations of culturally different people symbolically claiming *their* homeland. Thus issues of belonging and marginalisation osmotically slipped into our consciousness through the experience of living in a rural place. Likewise we saw poverty, alternative lifestyles, the search by in-migrants for different forms of rural idyll, and the social and geographical isolation of remote rural living for the first time. The idea that the nearest town of any size is a twenty-five-mile drive away (for those with private transport) and that for any major activity or purchase a 100-mile round trip to Swansea was necessary, quickly became the remarkable norm, frustrating some and seemingly not worrying others at all.

The experience of living in a small rural town, albeit with the potential 'shelter' of a university 'community within the community' undoubtedly coloured the interpretations of rurality and rural change contained within my writing. So too, however, did two other specific factors. First within a year of arriving there, two very influential characters also arrived – Mick Griffiths, a geographer from Reading, and Rob Young, an archaeologist from Durham. In that wonderfully close-knit work community that a place like Lampeter engenders, Mick and Rob unceremoniously demanded a more explicit articulation of political views than I had been prepared theretofore to divulge. Though neither was sympathetic to my Christian faith, they did encourage me to work out what that meant in terms of the political issues and the radical literatures of the day. Many would claim to have been influenced by the writings of Manuel Castells, David Harvey, Doreen Massey, Dick Peet and others in geography, but my somewhat late reading of these authors was embellished by discussions with Mick and Rob which seemingly went on for hours on end. Gradually I began to be more confident in articulating a socialist Christian viewpoint, which continues to be an influence on my geographical imagination fifteen years later.

The second influential factor I can only describe as 'college politics'. In a small college such as that at Lampeter, the lines of decision-making, influence and representation are much more visible than would normally be the case. The formation of micro-élites, the exercise of self-enhancing autocratic leadership, miserable attitudes towards the powerless, and the struggle for various forms of participatory democracy all seemed to be regular components of the everyday diet of action. Being an active consumer of this diet, and participant in some forms of representation within it, I began again to be influenced by some of these everyday ideas in my attempts to find concepts and theorisations which would inform interpretations of planning and policy.

This mix of lifestyle experience, shifting cultural and political consciousness and active political habitat meant that I began to beg new questions about rurality, rural change and rural planning in my research and writing. Much of my earlier work on key settlements and rural settlement planning had been based on the assumption that what was written in planning policy documents would be translated into equivalent actions in the places concerned. I began to doubt, however, that such simple translation was possible in the political world of policy-making, so I turned to the literatures produced by *implementation* theorists for guidance on the complex and interconnected webs of power and decision-making in rural policy-making and policy-enactment (see Cloke, 1987). Research which investigated the personalities, practices and politics involved in producing structure plan policies for the county of Gloucestershire, brought me into direct contact with planners and councillors, and was persuasive in leading to the suggestion that interpretations of local state activity cannot be divorced from the question of the roles played by the wider state itself. This in turn begged questions about the state, about different forms of power, and about the political economic nature of change (Cloke and Little, 1990). Despite the fact that mainstream literatures had been produced on these themes in urban and regional contexts, there was relatively little written about rural areas from a broad political economy perspective at this time.

Around this time, then, my writing about rural change and planning and policy in the state context took on a form of allegiance to the interpretative theme of political economy, which, as Peet and Thrift (1989) suggest, leads geographers to 'practise their discipline as part of a general, critical theory emphasising the social production of existence' (p. 3). Thus by the mid 1980s I was emphasising very familiar political economic themes in the telling of stories abut rural change:

> It is important to stress the centrality of *capital accumulation* as the driving force of social formation. The continuous process of reinvestment undertaken by individual capital units in search of surplus value is capable of generating unbalanced and unregulated trends of growth and decline. At various stages of this process deferrents to accumulation are experienced which require a restructuring of production to ensure a continuation of acceptable profit. Mechanisation, labour control, and

> market manipulation can be brought about without a spatial shift of production, but *capital restructuring* can also involve a relocation of production to a more favourable accumulation environment.
>
> (Cloke, 1989, p. 181)

The political economic perspective also took account of the processes of *social recomposition*:

> Clearly, restructuring does not take place in a societal vacuum. Different localities will have different histories of political and class conflict, will have experienced varying forms of social reproduction, and will exhibit particular contemporary class compositions. As such, class relations are not only the end-product of foregoing rounds of capital accumulation and restructuring, but also serve to mould the characteristics of ensuring iterations of these processes.
>
> (Cloke, 1989, p. 181)

and of the role of the state as a context for planning and policy:

> There are at least two levels of constraint experienced by policy makers with responsibility for rural areas. . . . First, the acceptance of an art of the possible which is conditioned by the overall state–society relationship presents decision-makers with an artificially narrow range of policy options – if social production and investment are designed to aid capital accumulation by the minority, and social consumption is confined by the need not to disturb the societal status quo, then it is the state–society constraint which underpins the continuing uneven distribution of power, wealth and opportunity. A secondary set of constraints arises from the complex inter-agency relations within and between the public and private sectors.
>
> (Cloke, 1989, p. 185)

In many ways, the adoption of a political economy perspective has very particular implications for the notion of rurality itself. It has been argued (Hoggart, 1990) that if the emphasis is to be on the structural changes in economy and society, then the category 'rural' becomes a rather unhelpful form of spatial delimitation. In Britain at least, many areas are culturally and physically urbanised to a greater or lesser extent, and the focal problems of powerlessness, poverty, social and cultural marginality, etc. occur across what is otherwise thought of as the urban–rural spectrum. If these arguments are accepted, it would be far better to undertake a series of comparative studies in different localities than to comply with seemingly artificial delimitations of urban and rural.

As can be imagined from the foregoing account of what it was like to live in a small remote west Wales town, I was never really able to accept that the 'rural' did not exist as a reasonably useful geographical category, despite the logic of arguments used by those of the opposite view. I did, however, begin to think and talk in terms of rural *geographies* rather than a rural geography so as to reflect the unevenness of change brought about in the context of these political economic trends. This is just one example of a number of uneasinesses I had with the forthright political economic accounts I was offering at the time.

Another involved the issue of reductionism which was (and is) a common criticism of these kinds of approaches. It seemed that however sensitively the political economy version of rural change was written, taking full account for example of the effects of locality, history of social composition and the like, the charge of being reductionist always had *some* validity in the non-compliance of individual places and communities to more general theoretical expectations. Yet another concerned the generalised assumption that political economic commentators could safely be labelled as Marxist – or at the very least neo-Marxist. Indeed I was saddled with this label more than once by reviewers, and yet in my own mind I had found considerable compatibility between political economic concepts and *Christian* principles and so was rather unhappy that the pathway to political economy was being perceived as uni-directional. And yet another concern was how issues such as gender and ethnicity sat alongside the political economy account. At the end of my chapter in *New Models in Geography*, I wrote:

> Recognising that class composition and political, economic, and cultural configurations in a particular locality can shape restructuring as well as be affected by it, there are important changes to be investigated resulting from particular iterations of the restructuring process. Changing class structures, particularly the infiltration of different fractions of the middle class (with marked impacts on local economies, political representation, and so on); changing gender divisions; and changing cultural characteristics represent just some of the major issues requiring serious attention here.
>
> (Cloke, 1989, p. 191)

This really did reflect some of the uneasinesses I was experiencing. Working with Jo Little on the 'Gloucestershire Structure Plan' project encouraged me to read some of the literatures being produced on gender; working with Nigel Thrift on the lead-up to our 'new middle classes' project led to a further consideration of literatures on class and other social divisions and a whole host of factors meant that 'changing cultural characteristics' became another 'strange' destination to be traversed to from the familiar interpretative 'home' of political economy.

IV

In considering how 'changing cultural characteristics' might usefully represent an important focus in the interpretative story-telling of the rural, I have more recently indulged in considerable self-reflection with regard to the appeal of the emphases on *structural* change within political economic approaches. After all, many others who have thoughtfully melded their Christian values with their geography or other social science appear to have been particularly concerned with matters of human agency and have therefore allied themselves to more humanistic traditions in the subject(s). In my own case a reading of biblical injunctions about how life might be lived leads me to *corporate* notions of 'good' and 'bad' action as well as to individual notions of the same. This links strongly then with a translation into the political arena which involves very strong views about the 'evils' of injustice, poverty, discrimination at the societal level as well as about individual ethics and moralities. I explain this, not to proselytise, but to account for a continuing fascination with changing political and economic structures as well as with the plight of the 'victims' of such structures. For example, to have lived and worked during the era of Margaret Thatcher (or to use an anagram I first saw used by Adrian Plass 'get rich team hag') has engendered in me a very strong (almost compulsive) wish to tell interpretative stories about the problems and victims of market-orientated (or self-orientated) programmes of government – deregulation, privatisation, the conversion of the welfare state into an 'ambulance state' and so on. The changes occurring in 'rural' Britain, and elsewhere where similar political programmes pertain, are to me inextricably bound up with these wider structural manoeuvres (see Cloke, 1992). This powerful personal entwining of Christian faith and broadly socialist politics has for me represented a huge barrier against the complete rejection of meta-narrative, at least in part because it does give me wide-ranging notions of 'truth' and 'good' which I would want to argue strongly within any pluralist setting.

All of this perhaps helps clarify (at least for me) why elements of critical theory and political economy remain strongly influential in my interpretations of the changing nature of rurality and rural areas. Nevertheless, other concerns have pushed their way into this crowded, and perhaps rather dogmatic intellectual space and imagination. First and foremost, I have to say that the steady stream of research interviews and less formal contact with people living in rural areas during the 1980s and early 1990s served continually to challenge as well as to reinforce the primacy of structural interpretations of change. It seems rather trite to relate that such interviews have been a constant reminder that life and lifestyle experiences (in this case in what are perceived of as rural areas) will more often confound that which is conceptually expected rather than confirm it, *IF* the story-teller is willing to allow herself or himself to be so confounded. This very basic idea of difference is ever present in these contacts with people and cannot fail to impress unless the researcher is too busy looking for sameness. This distinction may be one which most people undertaking research have grasped immediately, but I suspect there may be others who like me have closed off major potential areas of communication before interviews have even begun. This is not to suggest that the position is now one in which magically I see myself as having access to all available difference in the way in which I practise my geography, but rather that I have become a little more awake to the process whereby my self is an important influence in constructing text as well as in interpreting it. The door is also more open for the recognition of other geographies of the rural which my self would not immediately be in contact with even if using sensitive ethnographies as a means of practising research.

A second area of concern about political economy stories of the rural was prompted by a failure to reconcile ideas of economic restructuring, social recomposition, the role of the state and so on with my personal experience of living and working in a small town in west Wales. Here, partly because of the very obvious circulation of culture through the Welsh language, but also because of the very apparent differences for some in-migrants between the rural lifestyle of their imagination and their experience of a new life in rural Wales, the persuasiveness of *cultural* characteristics as influential and sometimes dominant elements of rural life could not be ignored. In the Lampeter area I met people pursuing a range of 'alternative' lifestyles: the back-to-the-land smallholder (and the sceptical reactions of local farmers to them); the travellers or more permanent dwellers living in what they described as the basic simplicity of, for example, an old vehicle or even a teepee (and the explosively negative reactions from most other sectors of society to these so-called 'hippies'); the in-migrant self-employed (often scratching out only a very basic economic living but often firmly believing in the compensatory value of 'rural' lifestyle) the middle-class often English, influx associated with the College (and so pidgeon-holed by many local Welsh-speaking folk as 'college people'); and so on. In each case, rurality was associated with particular cultural expectations, and cultural competences which resulted in a currency of ideas about 'belonging', 'marginalisation', 'theft', 'modernisation', 'naturalness', and so on and which were played out in the cultural realm. Moreover such expectations and competences crisscrossed with the exercise of power and can therefore be seen to promote conflict where different cultural idylls were incompatible in the places and lives concerned. All of this may just be a very long-winded way of saying that 'culture matters', but for me it has been these lived experiences as well as the more recent turns taken by social science intellectualism that have prompted an interest in the socio-cultural alongside the political economic.

Against this backcloth of life/research experience, I would also acknowledge the stimulus to reading new literatures and to addressing new philosophical issues presented by post-modern and post-structural thinking that was provided by my friend and erstwhile colleague Chris Philo. Looking back, his patient tutoring to 'think for myself' rather than overly relying on the existing thoughts of others has been an important impulse to self-reflection. His help in grappling with new literatures and his interested probing of the geographical imagination has made him an excellent teacher of teachers. Accordingly I have found that the very basic questionings that I have tried to describe over the last few pages are but rather rudimentary versions of long-standing and brilliantly argued debates in a range of historical and contemporary literatures. In delving into at least some of these I have discovered a (for me) very interesting tension between my personal wish to continue to grasp particular meta-narratives, and a growing sympathy with key ideas that are associated with those who by-and-large are rejecting meta-narrative: emphases on authorship, representation, discourse, ethnography, and in particular other geographies of the rural. I find Philo's (1992) arguments about 'neglected rural geographies' compelling in so far as my own male, white, protestant, middle-classness shields me from other stories of the rural relating to gender, ethnicity, class, sexuality, age and cultural proclivity. Equally, I find myself waiting to explore the interconnections between political economic power and cultural power as a way of interpreting some of these other stories.

In this account of what amounts to a shifting gaze on rurality, I have so far been fairly silent on the subject of rurality. Having admitted earlier on in this chapter that mechanistic statistical descriptions of rurality are usually self-defining and therefore of relatively limited use, and yet that the idea of 'doing away with the rural' as a useful category of study was somehow anathema to me, it does seem important to express some idea of the nature of 'country' that is being written. In one sense I baulk at this task because part of me does not wish to be identified with the institutional nature of rural geography (or rural anything) as a specific sub-discipline. There is a view that there has been a form of self-preservation amongst rural geographers who have been anxious to ensure the continuing survival of their field. Further, this has militated in many cases against the adoption of any form of theoretical stance which dares to suggest that social relations may be more important than spatial relations, and therefore that 'rural' may be an unimportant, or at best secondary classificatory device. Such a self-preservation has led to the rejection of critical theory 'heresies' which are seen as potential destroyers of the rural sub-discipline which has taken such strong institutional pressures to build up. Another problem with the task of conjuring up a 'version' of rurality is that my own interest in rural matters seems to stem from different stimuli than those of many other 'rural' specialists. Whether they have family connections with the agricultural sector, or whether it is their love for the countryside which has specifically kindled their interest, many 'rural' social scientists appear fundamentally attached to the extensive land use and landscape characteristics of rurality. Again, it seems likely that such attachments will prompt strong suspicions about any concept of the rural which does not prioritise the physical and functional nature of the countryside.

Despite these personal worries about the unhelpful complementarity between institutional pressures and somehow institutionalised definitions of the rural, my own lifestyle/research experiences do suggest to me that people do make decisions (where to live, where to set up business, where to go for recreation or leisure, etc.) which presume a category rural, whether this is imagined, or experienced in a material sense. Given this social construction of rurality, albeit one which will differ between individuals as well as be subject to wide-ranging discourses, there does appear to me to be scope to continue to tell stories which apply to the rural. Indeed if there is any advantage in seeking to place cultural characteristics alongside political economic concerns in an interpretation of people and spaces that are commonly recognised as being rural, then the idea of rurality as a socially and culturally constructed space may well be a most appropriate notion. It is certainly one which is beginning to be explored more sensitively. Mormont (1990) in discussing these issues concludes that it is no longer possible to conceive of a single rural space.

Rather, he points to a multiplicity of social spaces which can overlap the same geographical space, and thereby recognises rurality as a social construct with the 'rural' representing a world of social, moral and cultural values. The idea of rurality as a social construct, then, appears to offer rural researchers the opportunity to interrelate with very important contemporary themes in social science, as they express and practise an interest in the ways in which the meanings of rurality are constructed, negotiated and experienced. Clearly, such a search cannot be for *the* rural or for *the* idyll in rural life since it will be important to acknowledge a multiplicity of versions of rurality in policy, lay and academic discourses, with each version having different forms of social relations naturalised within them. Halfacree (1993) urges that the emphasis on social constructions of rurality in prompting academic discourses on rurality should be increasingly routed through lay discourses, thus emphasising the need to allow 'ordinary' people's own voices to be heard in, and to inform academic and policy debates. Again, though, it is worth stressing that there is no expectation here that people's constructions of rurality will all fit neatly together into a unitary thing called rural, since people are unlikely to hold clear, well-defined and well-structured images of the rural.

In accepting the idea of rurality as a social and cultural construct, drawing on individual imaginations and experiences of the rural which may reflect increasing distance between the sign and signification of 'rural' and their locational referent, my own research interests have represented the taking of small steps towards interpreting these interwoven strings of constructing rurality, principally by drawing together in different ways, elements of political economic theorisation with elements of the symbolic importance of notions of rural idylls. I do not claim any originality for this work – indeed it has involved a return to previous literatures, for example to historical studies of rural community as well as to a re-reading of more recent writings of say Raymond Williams and Stuart Hall. It is, more correctly, what fascinates my own geographical imagination at present, and although this in turn raises the important question of why this fascination should be spoken out in the public domain, I will leave that question until the end of this chapter, and perhaps unanswered altogether.

What follows here are four stories of how elements of political economy and cultural constructs of rurality have been interlaced in some of my recent writings. These are not in any way meant to represent the 'only' ways, the 'best' ways or even 'good' ways of (en)culturing political economy. They are merely the first stories to have emerged for me, given the particular circumstances of my self and my research strategy over the last two or three years.

V STORY 1: STRUCTURED COHERENCE OR UNSTRUCTURED INCOHERENCE?

So much of the readings and writings within rural geography as in most other subject areas stem from very particular cross-fertilisations of literatures, presentations, discussions and contacts with the researched, and these four specific interpretative stories are no exception. The arrival of Mark Goodwin in Lampeter heralded a very fruitful period of discussion and research collaboration for myself and others. Mark's work on post-fordism and changing urban governance (see, for example, Goodwin *et al.*, 1993) seemed to cover, in much more polished and complete fashion, some of the conceptual ground which I had been attempting to apply to the rural context. In particular I had been puzzling over how, if at all, some of the middle-ground concepts arising from the regulationist school(s) made sense of the various rounds of economic restructuring and social recomposition which appeared to have been occurring in the rural domains of western countries. Between us Mark and I began to work through the ideas of mode of regulation, societalisation and structured coherence in relation to rural change (Cloke and Goodwin, 1992).

The concept of a *mode of regulation* suggests that there is an ensemble of institutional forms, networks and norms which works to ensure the reasonable compatibility of behaviours in periods of stability between production and consumption. It follows that any change in regulation, whether by state intervention or by means of social struggle and opposition will in turn lead to changes in the experiences of rural places and lifestyles by rural people. The changing mode of regulation therefore perhaps offers to link changes in rural production with changes in the living and thinking and feeling of life in rural areas. *Societalisation* further suggests that there is a process of regulation at a societal level, with a complex ensemble of social practices which operate to integrate diverse social structures and to secure some form of cohesion among competing forces. Jessop (1990) has talked about the recognition of both historic blocs (a historically constituted and socially reproduced correspondence between the economic base and the political and ideological structures of social formation) and hegemonic blocs (durable alliance of class forces who exercise political, intellectual and moral leadership) in the operation of societalisation. It is therefore possible that particular forms of societalisation in rural areas will influence social, political and cultural leadership, including the circulation of dominant constructs of rurality. Mark and I took the view that currently in the UK there is an ongoing appropriation of cultural values from previous historic and hegemonic blocs in order to promote a commodification of the countryside which itself underpins the emergence of new blocs.

The third middle-ground regulationist concept we used in these interpretations is that of structured coherence which is drawn from the work of David Harvey (1985). What we wanted to do was to try to find a way of describing the way in which places form the essential meeting points of changing mode of regulation and societalisation, and our reading of Harvey's idea was that it could help us to look at the interlinking of mode of regulation and societalisation through particular relations and institutions which apply to particular places at particular times. Any coherence which was evident would be structured not only by the prevailing form of production, but also by innovations in standard of living, lifestyle, social hierarchies and sociological and psychological attitudes towards working, living, enjoying, entertaining and the like.

When we first unveiled these thoughts in a seminar, the immediate reaction was that we had reverted to (or remained in) the metatheoretical mode, offering little concession to matters of cultural construct of rurality, or cultural experiences of rurality. It was read as just the same old political economy by those present, even though it was written with different intent. We were certainly not suggesting in this some form of overarching grand theory which could instrumentally dictate the structural conditions under which places always achieve a recognisable coherence at any particular time. Indeed we started from the premise that contemporary society is messy and complex in nature, yet we felt that rural places could be categorised (indeed are so categorised by some of their residents and visitors) according to the specificity of place and people without divorcing that place and those people from wider sets of changing relations. I suppose my own wish was to begin to discuss new sub-regional rural geographies of particular places, whereby localised coherence emerges in the midst of partial and contested temporary stability, and within which it would still be necessary to recognise differential experiences and imaginations of localised coherence. Thus different types of coherence will be differentially attractive to various forms of capital accumulation, to those seeking a rural experience, and to state intervention.

In working through these three processes of restructuring, recomposition and state intervention (Cloke and Goodwin, 1993) we were anxious to address the question of whether it is possible to interconnect political economic concepts with the idea of socially constructed attitudes and behaviours. Certainly we found that when discussing rural change, those components relating to state activity and to processes of accumulation and exploitation by capital are readily compatible. Yet it was our view that the cultural issues of the 'attractiveness of a place' and of the availability of a rural idyll were of equal importance as potential components of rural change. These issues appear vitally interconnected to processes of restructuring and recomposition and seemed to us to be extremely important in the characterisation of particular rural places, whether they change or remain the same. Therefore, although it is important to stress that the changing functions of rural areas are by no means uniform or predictable and that it is certainly important to avoid overgeneralisation, there do seem to be some grounds for suggesting that the idea of socially constructed rurality need not be incompatible with concepts which locate specific places and people in wider sets of changing relations. Indeed we concluded that it may be very important that the contracts and strategies of capital in altering institutional forms, networks and norms, the impact of the contesting of change in socio-political spheres, and the role of cultural factors as a kind of glue in establishing locally coherent characteristics are brought together in our analyses rather than being regarded as belonging to separate philosophical domains.

It is always difficult to know how a particular piece of story-telling is received by different readers. People who we feel that we can trust to give honest answers to this have presented readings which differ enormously. Some readers (including an editor and referees) thought that there was a useful message in this attempt to (en)culture particular elements of political economic thinking. Others suggested that all we had done was to present some very obvious and well-known ideas and dressed them up in fancy conceptual language – a form of intellectual obfuscation of interpretation designed to appeal to restricted élite audiences who shared the common code of jargonised text. Yet others were worried that we had allowed far too much political economic theory to dominate the outcome; that *any* talk of structure or coherence should be immediately deconstructed, and that the rather more appealing idea of unstructured incoherence be firmly instated as the basis for interpreting cultural notions of rurality. My own inability to dispose of meta-narrative (and indeed Mark's own different positionality on this point) appears to be an obstacle to communication in some of these latter cases.

VI STORY 2: COMMODIFIED RURAL SPACES

For as long as I can remember, I have made a practice of calling in on Tourist Information Centres when I was in an area away from home, whether on holiday or not. I don't seem to be alone in this practice because these places seem invariably to be busy with people picking up ever more appealing brochures outlining the holiday/leisure/recreation attractions available in the area. The centres have become meeting places between leisure consumers and commodified spaces for leisure. My impression is that increasing numbers of people spend increasing amounts of time visiting these commodified places, and I have a fascination about how the proprietors of such places represent them in their literature. It was in this practice of 'collecting' brochures, and taking an interest in these places that I began to think about them as examples of wider processes of commodification both in rural areas and of rurality.

In a chapter describing 'Thatcher's countryside', I commented on this seemingly significant shift in the nature and pace of commodification in rural Britain which was giving rise to a series of new markets for countryside commodities:

> The countryside as an exclusive place to be lived in; rural communities as a context to be bought and sold; rural lifestyles which can be colonized; icons of rural culture which can be crafted, packaged and marketed; rural landscapes with a new range of potential from 'pay-as-you-enter' national parks, to sites for the theme park explosion; rural production ranging from newly commodified food to the output of industrial plants whose potential or active pollutive extremities have driven them from more urban localities.
>
> (Cloke, 1992, p. 293)

It seems to me that wrapped up in these broad processes of commodification and in particular in the various representations of specific commodified places, there are fascinating illustrations of how rurality is being constructed by a relevant group of interests. Through this rather odd series of interests and circumstances I was able to practise in a small way some of the broad literatures on commodity, sign and signification in the context of rurality. Seeking a framework for this task, and also I think seeking some confirmation of wider (and difficult, at least for me) reading of, for example Guy Debord and Jean Baudrillard, I was grateful to Clare Fisher, a postgraduate at Lampeter, for sharing with me an excellent paper by Best (1989) which traces the conceptualisation of commodification through a number of different 'societies'.

For example in dealing with what Marx wrote about commodification, Best identifies the point at which an object assumes an exchange value which is over and above its use value as the point of commodity. In this way commodification denotes an abstraction of the object from its use value and therefore from social references about the need for the object and the quality of the object. A commodity thus takes on judgements of worth which are socially and culturally constructed, and Best terms this *the society of the commodity*. Debord advances this idea into the context of late capitalist society, seeing social control as being achieved by consensus rather than force. This concensus is based on a cultural hegemony realised through the transformation of the society of the commodity into the *society of the spectacle* – a world in which consumption is constructed by others and which consists of a series of spectacles designed to pacify and depoliticise. As Best (1989) remarks, 'The spectacular society spreads its narcotics through the cultural mechanisms of leisure and consumption services and entertainment, within a culture that has grown (relatively) autonomous from the social totality' (p. 29). This view would suggest that society is geared towards the production of spectacle and that the social and cultural constructs of commodity (including in this context commodities of the rural or the countryside) should be seen in that light; commodification for spectacle is explicitly concerned with the production of illusive and artificial counterfeits of real objects and relations.

Best then turns to the work of Baudrillard (especially 1983a, 1983b) to suggest another important stage in the commodification of reality:

> With Baudrillard, we move to a whole new era of social development . . . We leave behind the society of the commodity and its stable supports; transcend the society of the spectacle and its dissembling masks, and enter the society of the simulacrum, and abstract non-society devoid of cohesive relations, social meaning, and collective representations, an imploded socius of signs.
>
> (Best, 1989, p. 33)

He characterises Baudrillard's descriptions of post-modern cybernetic society as the *society of the simulacrum*, in which exchange is carried out at the level of signs, images and information, and in which commodification is in an abstract sense the absorption of the object into the image so as to allow exchange to take place in semiotic form. This society is replete with sign-exchange values, and commodification is intertwined with the structural logic of the signs concerned. Production and consumption of commodity thereby focus on the conspicuous nature of social meaning and are translated into a system of abstract signifiers which are not tied to any necessary relationship with the objects of commodification. Thus for Baudrillard, the commodity is eclipsed by the sign, which can be unrelated to the reality of the commodified object altogether.

For me, a very specific small research question proved to be the gateway to much wider issues about rurality, spectacle and simulacrum. The small question involved the tourist and leisure attractions which have become so much a part of the visitor information culture. To what extent were these represented as being 'rural' spaces and places, and what elements of spectacle and simulacrum were evident in the symbols used to represent and to attract? In a very preliminary discussion of these questions, using representations and iconographies associated with 'rural' attractions in Wales and south-west England (Cloke, 1993), I concluded that the society of the commodity did appear to be firmly rooted in the countryside and that this was so at least in part in response to wider political economic pressures from privatising and deregulating policies from the state which have opened up the opportunity for specific spaces in the countryside to be developed as pay-as-you-enter attractions to tourists and leisure-seekers. This is not, of course, a new phenomenon of countryside recreation, but the speed and scale of change does contrast with previous conventions about leisure and recreation in the countryside which stressed a 'get away from it all' experience in which access to the natural is free.

Similarly, a detailed exploration of what is on offer (and equally important, how what is on offer is represented to potential consumers) does allow the suggestion that a kind of society of the spectacle is also emerging in the contemporary British countryside. The brochures for these attractions will very often claim that a visit will allow a 'rural', 'countryside' or 'natural' experience, and they do often signal that the social and cultural constructs of a commodified countryside are fashioned in terms of the production of spectacle. Sometimes this is very specific, as in the case of Morwellham Quay in Devon, which offers an outdoor theatre of the rural history. Its brochure issues a series of invitations for visitors to participate and spectate in captured history:

> the quay workers, cooper, blacksmith, assayer and servant girls dressed in period costume, recreate the bustling boom years of the 1860's. . . . Chat with the people of the past. Sample for yourself the life of the port where a bygone age is captured in the crafts and costumes of the 1860's. . . . Try on costumes from our 1860's wardrobe.

Many of the attractions I looked at offered spectacle of one form or another. The National Shire Horse Centre in Devon, for instance, presents a working farm in which heavy horses rather than modern machinery are used to carry out the heavy duties of farm work. This is now a major tourist attraction for visitors to the Torbay and Plymouth areas, and displays of shire horses and rides on waggons pulled by shire horses reinforce the specific theme of the attraction. There are, however, special events at the Centre which supplement the theme of farming in a bygone age. These include western weekends (the brochure claims 'you will see Cowboys, Indians, Outlaws . . . with an authentic shoot-out display') 'Teddy Bear's' Picnics and 'Dolly's' Tea Parties; steam and vintage rallies and classic car shows. Many other attractions also offered this mix of the traditional rural and other symbolic heritages as part of their representation of the rural history of a rural space. Perhaps the most striking example of spectacle in these brochures was the way in which historic sites are being commodified using symbolic events:

> Cadw in Wales, for example, now has a programme of theoretical, musical and other performance events at their Castles, Abbeys and Palaces. Witness living history with displays of archery, armour, combat and dancing at Raglan Castle; marvel at huge and spectacular kites representing the red and white dragons from Celtic mythology at Beauman's Castle; Tretower Court is the 'perfect backdrop' for a celebration involving 'theatre, romance, music, crafts . . . and even medieval alchemy'; listen to the Llannig Silver Band amongst the museum workshops and quarries of the Welsh State Centre at Llanberis.
>
> (Cloke, 1993, p. 65)

The message which I drew from these representations of attractions was that the countryside is being commodified in such a way as to go well beyond the real objects and relations of the places, sites and buildings concerned. All of the attractions which I looked at stressed their rural or countryside habitus. Some offered spectacle that made claims towards authenticity in that habitus, particularly where attempts were being made to recreate history. Elsewhere, however, the conspicuous consumption of the symbols which were being offered to visitors – western shoot-outs, vintage cars, teddy-bear's and dolly's tea parties (with the obvious gendering of symbols there), craft, alchemy, etc. – suggests that some rural attractions are using representations and iconographies which are related neither to the particular place, site or building concerned nor to its landscape or history.

In pursuing this small question of whether representations of rural attractions indicate signs of spectacle or simulacrum, the larger question of rurality in contemporary society also becomes for me an important focus. Rurality as a social and cultural construct is subject to a constant flux of production, consumption, reproduction, representation, commodication, manipulation and so on. The interplay between power and the cultural realm is for me as vital a subject of interest as is the perhaps more obvious, or at least more conventional interplay between power and the material political economic realm. Ruralities are being made and remade, experienced and re-experienced. Rural life reflects at one and the same time the boundlessness of the imagined landscape and community and the restrictiveness of access to the material and cultural conditions which permit the imagined to be lived out other than in imagined form. The notions of power both over the imagination and over material and cultural conditions remain crucial for me in reaching an understanding of rurality and rural life.

VII STORY 3: SOCIAL AND CULTURAL CONSTRUCTIONS OF RURAL POVERTY

I suppose that my Christian and political convictions weld together particularly strongly on issues such as poverty. I align myself very easily with Old Testament prophets like Amos, ranting frequently against the modern-day 'cows of Basham', and there is no doubt that my personal geographical imagination is fired by a wish to understand, publicise and address issues of poverty, starvation, famine, and so on wherever they occur in the world. I have found the increasing tendency during the 1980s of 'blaming the victim' in these areas totally abhorrent, and I find it very difficult to retain a degree of personal self-control when governments appear blatantly to ignore, hide or devalue issues of poverty just by talking about them (or not as the case may be) in particular 'clever' ways.

Having identified my own prejudices, I must also admit to some puzzlement when attempting to identify the exact nature of the 'problematic' of rural lifestyles in Britain and other similiar situations. Although in global terms, the relative poverty experienced in Britain's rural areas can easily be downplayed, nevertheless my own lifestyle in rural Wales, and the specific excursions I have made to other rural areas in order to interview people on such matters, suggest to me that there *is* an often hidden 'problem' of poverty in rural Britain. Yet these experiences equally suggest to me not only that because I am who I am, I will not have 'contacted' the problematic experiences of many different individuals and groups living in rural areas, but also that those who I would think of as experiencing poverty or other such difficulties will frequently not wish to be so represented.[2] My idea of their 'problematic' does not accord with theirs. So while we merrily talk, write and read about 'rural deprivation' I have the uncomfortable feeling that we don't really know with any precision what we're talking, writing and reading about. At the very least, I have long felt that an understanding of rural deprivation and poverty necessitates not only an interpretation of rural lifestyles, but also an interpretation of the circumstances in which rural lifestyles are experienced as problematic.

Such a realisation that experience of rural life is highly differentiated is, of course, not novel. In the rural studies literature for example, Knox and Cottam (1981) have stressed the importance of 'People's *effective* response to the environment: feelings which may intensify or compensate for "objectively" defined deprivation' (p. 173), and Newby (1987a) has identified the lack of consensus about satisfaction with 'community spirit' in rural areas: 'Should account be taken of only the material standards of village life, or should more

[2] Some interesting empirical evidence on these points from the Lifestyles in Rural England research project is presented by Woodward (1992).

decisive qualities, such as the sense of identity and belonging, be included? And from whose perspective should any accessment of gains or losses be taken?' (p. 225). However, the *practice* of a geography of rural poverty or other 'problematics' which acknowledges both the unevenness of material opportunities and the varying social and cultural constructions of rurality, community, living standards and welfare, has not been nearly as common as these recognitions of the need to account for both the material opportunities and the social and cultural constructions of rural life.

My own opportunity to research these issues has come with funds from various government agencies to carry out a study of 'rural lifestyles' but the degree to which I wanted to incorporate culturally constructed notions of the rural problematic in this research can only be understood by acknowledging the impact of a three-month study visit to the USA in 1990 sponsored by the Arkleton Trust. This provided me with an opportunity not only to visit different parts of America with normatively high levels of rural poverty, but also to do some uninterrupted reading of American literatures on poverty, comparing these with those dealing with the British case. It was here that I began to accept that the experience of marginalisation, when personal or minority constructs of rurality, community, living standards and welfare are overpowered by dominant constructs in these areas, may be as crucial to an understanding of the problematic of rural lifestyles as the previously emphasised changes in material circumstances. I have been particularly struck in this regard by the arguments of Handler and Hasenfeld (1991) in their book *The Moral Construction of Poverty* (which in fact I read after returning from the USA). They suggested to me that work on changing modes of regulation (see Section V) needed to be conjoined by rather more subtle acknowledgement of the way in which changing political economic conditions have been accompanied by particular discourses on welfare in which particular social and cultural constructs of poverty are highlighted, others appropriated politically, and yet others denied any existence in the semiotics of government. They argue that:

> Social welfare policy cannot be fully understood without recognising that it is fundamentally a set of symbols that try to differentiate between the deserving and the undeserving poor in order to uphold such dominant values as the work ethic and family, gender, race and ethnic relations.
>
> (Handler and Hasenfeld, 1991, p. 11)

Again, this emphasis on the discursive and the symbolic points to another way in which the political economy of rural welfare, and the cultural construction of rurality and poverty can converge. Changing welfare policies seek to make a series of symbolic statements to wider society about the codes of behaviour which are deemed to be appropriate or disfunctional. However, the rural order underlying these symbols of welfare is conflicting and contested, and therefore the symbols themselves can be ambiguous, and resultant state actions can be somewhat contradictory. Therefore, one of the major functions of changes to the welfare state will be to confirm, and if necessary shift, the dominant cultural norms about poverty.

In pursuit of these themes, and using these literatures, I have been trying to think through the idea of a discursive transformation of rural poverty and the welfare state in Britain and the USA (Cloke, 1994). In so doing I have accepted that the politicisation of particular social and cultural constructs is important in changing state attitudes to welfare and poverty. I do not see this as divorced from more political economic accounts of the changing welfare state – rather that part of the ensemble of economic, political, social and cultural relations to be formed around new modes of regulation has been the redefinition and reformulation of welfarism at state level. Although there have been clear differences in the USA and Britain over the redefining of welfare (in the USA for example there is an institutionally defined poverty line, whereas in Britain the manoeuvre has been to deny the existence of poverty altogether) there does seem to have been a common practice of using changing ideas and discourses of welfare and poverty to make symbolic statements to the wider society, essentially formulating clear and public notions of what is virtuous and what is delinquent behaviour. The heightened emphasis on dependency, for example, may be identified in the continuing recourse to the discourses of 'culture of poverty' and 'underclass', both of which I would argue have been appropriated politically and used to identify a discursive repository for the discrimination against and the exclusion of the 'undeserving poor' from wider society. Work in Britain, for example by Dean (1991), Mack and Lansley (1985), and Oppenheim (1990), clearly illustrates that through a series of contested transformations, the dominance of governmental discourse has bought about a redefinition of both the community and the individual, and of the way in which there is a relationship of responsibility between one and the other.

One question of particular interest to me concerns the nature of the social powers which have underpinned this discursive transformation. Ideological programmes from governments have to be seen in conjunction with the material changes associated with the rolling back of the welfare state during what has been described as the post-fordist era. In Britain, then, the discursive transformation of poverty and welfare was not contained within the apparatus of the state, neither was it a solely Thatcherite phenomenon. Stuart Hall has addressed this issue:

> The discursive relations of power cannot be constituted exclusively on the terrain of the state. They precisely crisscross the social body. There is no moment in which the powers that cohere in the state can ever exhaustively resume those that are dispersed through the plurality of practices in society. Thatcherism, as a discursive formation, has remained a plurality of discourses – about the family, the economy, national identity, morality, crime, law, women, human nature. But precisely a certain unity has been constituted out of this diversity.
>
> (Hall, 1988, p. 53)

It seems to me that much more work needs to be done to gain appropriate understandings of how discursive norms are reproduced and shifted in these kinds of areas.

The overriding relevance for my own research of this idea of a discursive transformation of welfare and poverty relates to the potential role of rurality in these wider social and cultural constructs. Drawing any urban–rural distinction here has to be prefaced with important caveats which convey to the reader that I do not conceive of rural areas as being in any way homogeneous, neither do I suggest that the perceived idylls of rural lifestyles in one nation will be replicated elsewhere (see Section VIII). However, in my reading on the transformation of dominant discourses concerning poverty and welfare, it did emerge that society-wide constructs do seem to differentiate between the urban and the rural. In the USA, rural poverty is symbolised as deserving rather than undeserving. In Britain, the already poor levels of public-sector welfare services allow rural areas to be symbolised as entirely compatible with new political cultures of privatised welfare. Moreover, in Britain it can be suggested that dominant social constructions of rurality may include cultural notions of idyll which render rural poverty, rural deprivation and rural disadvantage basically as contradictions in terms. These constructs seem likely to be major influences on the experiences of individuals as they encounter and construct 'problematics' in their rural lifestyles. The influence of idyll may be positive, as more than compensation for lack of material opportunity, or negative, as the non-experience of expected idyll serves to add to the experience of material hardship.

VIII STORY 4: RURAL IDYLLS?

The discursive issues raised both in reading literatures on poverty and welfare, as well as in living in the Thatcher's Britain of the 1980s, seem to point to cultural constructions of rural idylls as a potentially important meeting ground for studying experiences and opportunities in rural lifestyles. The notion of idyll has become an all important questioning logo in the research I have been involved with over the last five or so years. How are new middle-class residents in villages responding to different cultural idylls in their decision to move, their decision to furnish, their decision to participate in local community life, and so on? What is the precise nature of these idylls? How are they circulated and reproduced? Further, in seeking to interpret the potentially problematic nature of rural lifestyles, is the notion of 'problem' being undermined because particular representations of what rural life is like and should be like are dominant in the minds of relevant politicians, professionals and rural dwellers? Does the dominant representation of rural idyll feature happy, healthy and problem-free images of a rural life safely nestling within both a close social community and a contiguous natural environment?

If a dominant idyll does come to define the cultural domain of a rural place or area, then clearly ideas of deprivation and disadvantage *can* become a contradiction in terms. Such a contradiction, as Paul Milbourne and I have stressed, applies

> Both for those who experience hardship (but will perhaps see this as an acceptable trade-off for the benefit of rural living, or will seek for some reason to conceal or underplay the stigmatic acknowledgement of hardship) and for those who do not (and perhaps are anxious to reproduce the culture of an idyll by playing down any hardship that comes to their attention).
>
> (Cloke and Milbourne, 1992, p. 359)

As we also suggest, the idea of a rural idyll appears to be at best rather speculative at this stage: 'Much more needs to be known about the degree to which it is important in representations of the rural; the varying nature of idyll; the relative significance of "pro-rural" factors as opposed to anti-urban factors; and so on' (Cloke and Milbourne, 1992, p. 359).

There is already, of course, an emergent literature on these themes (see, for example, James, 1991; Mingay, 1989a; Short, B., 1992; Short, J., 1991) but during the rural lifestyles research projects (and particularly that based in Wales) I have become increasingly interested in unpacking the notion of different experiences of different rural idyll(s). It is surely far too simplistic to attempt to understand the nature of rural life in a particular village in terms of one composite and commonly held construct of rural idyll. Surely, the everyday processes of in-migration, and of the 'sedimenting' in of different rounds of colonisers into a particular place mean that different versions of idyll are continually being brought to that place, and ageing with the experience of living there. Equally surely, there will be substantial differences, both geographically and socially to the idylls that individuals carry with them, and in the cultural competences required of them for the process of 'belonging' to a more corporate idyll. I start then from the assumption that different cultural constructs of the rural are represented in the modern countryside and that these interconnect in complex fashion in a dynamic world where real social processes both underline and contest dominant cultures.

It is often national-level ideologies which are viewed as the conventional focus for the source of meanings about rurality, the circulations of meanings and the power relations vested in those meanings. However, it can also be suggested that the regional and local scales are also arenas in which important meanings about rurality are circulated and negotiated. This is not to imply that the identification of different scales of cultural circulation will lead to any functional, mechanistic or deterministic view of rurality and culture, as cultural constructs of the rural will be complex and defy neat categorisation. Nevertheless, within any particular rural space different cultural constructs of rurality combining different scales of received and circulated meanings will be evident and so the issue of scale may be a useful starting point for interpreting variations of idyll and for recognising different levels of power relations associated with cultures of rurality.

At a national level, I have become satisfied that there is some evidence for a series of circulated meanings associated with a rural idyll drawing on the settlements and landscapes of a mythically timeless and natural England. These meanings, though very difficult to pin down, point to rurality as an 'other' world – bucolic, problem-free, natural, happy and healthy. It is rooted in pastoral images of agricultural life, and augmented by the supposed sins of urbanism. It can represent arcadia, utopia, or a refuge from modernity (Short, 1991), both a nostalgic return to the natural roots of the nation and a futuristic colonisation of suitable space in which to practise adventitious consumption. Nairn (1977) suggests that the countryside has become part of the national identity of England; an expression of the natural homeland or territory.

I enjoyed reading Mingay's (1989b, 1989c) collections of essays which show how particular facets of historical rural life have become converted into idyll, sometimes through ignorance or sheer romanticism, sometimes as a continuing statement of power and credentials by dominant socio-economic groups, and usually making myth out of hardship and sanitising the unacceptable. The interconnections between past cultural paraphernalia and contemporary life are both material and symbolic. For example the country houses and country towns of the thirteenth century provide valuable heritage capital and theatres of consumption for service class colonisers of the late twentieth century (see Cloke and Thrift, 1990), while they also provide an arena for buying into the *values* of deep England. In this and other instances, the rural idyll can be bought and sold, such that power and wealth underlie the ideology of the idyll. Elsewhere, the same cultural notion of the idyll is used to sell other unrelated products, as advertisers turn to rural logos in order to appeal to idyll-laden consumption.

These national-level meanings are circulated in a number of ways (James, 1991) including art, literature, television and radio, advertising, newspapers, magazines, academic texts, as well as being promulgated through 'country' organisations, the designers of homes, home furnishings, clothes and other products in which idyll can be commodified as taste or style. There is a sense in which the idyll-ised view of village England does represent a backcloth to the expectations and aspirations of rural lifestyles – so much so that country dialects, dances and festivals have become entertainments demanded by the new gentry for whom 'authenticity' and theatricality are often synomyous, and on behalf of whom long-standing residents are called on to be actors and stagehands and farmers are expected to be 'scenechangers' (Newby, 1987b). There is also a sense in which the conflicts and negotiations over these broad representations of the rural will directly affect people's experience of rural lifestyles.

There has been far less discussion of the construction and circulation of meanings of rurality at the regional scale, and yet again I feel that the contemporary rural world is replete with messages about the cultural regionalisation of rurality. Gilbert's (1988) discussion of new regional geographies suggests that regional specificity can arise as a local response to capitalist processes, as a focus of cultural identity and as a medium for social interaction. When seeking to understand cultural constructs of the rural at a regional level, the second and third of these regional specificities are obviously relevant to the rural arena, and may be read into the first. Most of the illustrative material of these new regional geographies – and their interesting interplays between the political economic and the cultural – have focused on urban-industrial regions, and thereby regionalisation of rurality has usually been understood as a contrast to the urban-industrial nature of other regions. As Paul Milbourne and I have pointed out,

> Even Gramsci (1971) in his accounts of regional differentiation in Italy uses rural regions as a foil to the important statements he makes about city regions. Thus the spatial and social relations of subordinating a labour force are seen to lead to a quietly compliant, rurally based group of manual workers in the South of Italy in contrast to the radical middle management which was noted in particular Northern cities. This rural peasantry lacked the dynamism and innovation of its counterpart in urban regions because its intellectuals were drawn in large numbers into the clergy so reproducing the means of their own class subordination.
> (Cloke and Milbourne, 1992, pp. 362–3)

But as Cooke (1985) emphasises, 'It is quite clear that rural social relations can be antagonistic and non-deferential, through Gramsci tends not to make much of this important point' (p. 217). Rural regions may well represent the spatial organisation of particular social processes associated with the mode of production, social divisions of labour and networks of accumulation and political domination, but I feel that it is important to temper any potential reductionism here with a recognition that it will be the *interconnections* between identity, interaction and political economy which will characterise regional ruralities.

Many such interconnections are to be found in the narratives of individuals about the (rural) regions in which they live, such as Winifred Foley's (1974) evocative account of her childhood in the Forest of Dean:

> Ten by twenty miles of secluded, hilly country; ancient woods of oak and fern; and among them small coal mines, small market towns, villages and farms. We were content to be a race apart, made up mostly of families who had lived in the forest for generations, sharing the same handful of surnames, and speaking a dialect quite distinct from any other.
> (Foley, 1974, p. 13)

This is the kind of recollection often associated with 'days gone by' approaches to rurality, and it can further be implied that with the advent of modern technology acting to bring about a compression of time and space such regionalisations have long since disappeared. I think that there are at least two reasons, however, for regarding regional constructs of rurality as being of continuing importance. First, it does seem that cultural characteristics of regions continue to figure significantly in the social consciousness and social relations of groups of people in particular places. This can be seen in Cohenesque anthropological studies of belonging (for example, Cohen, 1983), and in more general recognitions that the cultural geographies of regions *do matter*, for example Doreen Massey's (1988) account of the changing images of the north of England, where 'There is less mention of satanic mills. More, the talk (in the south) is of how *wonderful* the countryside is, and the quality of life it is possible to have, and of how low house prices are' (p. 17).

I am working on the premise, then, that traditional regional constructs of idyll-ic rural life are being reproduced, reinforced and re-presented both through the reflexive experiences of new in-migrant groups in the areas concerned and through the mechanisms by which regions are marketed as a commodity, particularly as tourist and leisure destinations. In this latter context there has been a recent surge of specifying regional identities – Herriot Country, Bronte Country, Hardy's Wessex, Wordsworth Country, Robin Hood Country and so on – according to commodified mix of literary myth and idyll.

The second reason for regarding regional constructs of rurality as being of continuing importance is the suggestion that particular rural regions are the foci for political and cultural struggle. There are, of course, longstanding claims for regional sovreignty in particular English regions such as Cornwall, but to these I would add accounts of the reproduction and revival of cultural platforms in the politics of particular rural regions. For example, Keating (1988) describes the political, linguistic and literary revivals associated with the Breton movement in France, where the political manifests of regional identity have been accompanied by particular movements of cultural identity, such as the music of Alain Stivell whose folk-rock mix of popular protest songs, Breton musical traditions and pan-celtic cadences have become symbolic of the region itself.

In identifying a specific regional scale at which constructs of rurality are constructed, consolidated and reproduced, there is obvious scope for the recognition of tensions and discontinuities both between nationally circulated constructs and regionally circulated constructs, and between different regional constructs. Indeed some attempts to promote a specific regional construction of the rural (or 'a country') may well involve attempts either to undermine dominant national constructs, or to compete with alternative regional constructs in the various discursive contexts over rurality.

I believe that local-level constructs of rurality contribute further to the cultural mêlée of idyll, representing particular engagements with the experiences of specific rural places at the local scale. Such experiences then colour the reflexive communications made to others about rurality in more general terms. I make no apology for illustrating this point with a passage of text from the writings of Raymond Williams (1973) which I have used before in this context. In the piece that follows I see the man who has been so influential in debates about the culture of country and city revealing a very personal portfolio of meanings about local rural life, in which the writer as both voyeur and participant reveals experiential meanings associated with the nature, society, sight, smell and taste of country living:

> It is the elms, the may, the white horse, in the field beyond the window where I am writing. It is the men in the November evening, walking back from pruning, and their hands in the pockets of their khaki coats; and the women in headscarves, outside their cottages, waiting for the blue bus that will take them, inside school hours, to work in the harvest. It is the tractor on the road, leaving its tracks of serrated pressed mud; the light in the small hours, in the pig farm across the road, in the crisis of a litter; the slow brown van met at the difficult corner, with the crowded sheep jammed to its slatted sides; the heavy smell, on still evenings, of the silage ricks fed with molasses. It is also the sour land, on the thick boulder clay, not far up the road, that is selling for housing, for a speculative development at twelve thousand pounds an acre.
>
> (Williams, 1973, p. 3)

Local constructions of rurality will not be ubiquitious, and will be associated with different experiences of rurality and with different strategies for living in, coping with, and adapting to rural places – a point to which I return at the end of this chapter.

For me, at least part of the point of recognising different scales at which ruralities and idylls are constructed and circulated, is to address the question of how power is exerted through such constructs particularly how a dominant form of culture idyll is upheld, comes to be challenged, and perhaps is reproduced in a different form (see Cloke and Milbourne, 1992, for a discussion of this). Dominant idyll – related cultures of rurality currently appear to marginalise particular social groups and thereby incorporate discriminatory practices – ethnic prejudice, homophobia, anti-'hippy' or 'traveller', and the like. The village can often represent a cultural arena of conflict, and therefore lifestyles in that village will often be directly affected by cultural motifs of (to use words commonly used to us in our interview responses) 'attack', 'defence', 'resistance', or 'counter-attack'. Such motifs can permeate all aspects of life, and particular rural problems experienced in these places can often be interpreted in terms of cultural defence or resistance. This in turn informs cultural motions of who is *welcome*, or who *belongs* in a village, thereby reproducing and reinforcing processes of socio-cultural marginalisation. It is therefore very important to recognise the overlapping and interconnecting of diverse and often conflicting cultural constructs of rurality in a particular place or region. Some oppositions will be at the symbolic level, whereby discord between different lifestyles in the same place may be brought about by the lack of symbolically crucial *cultural competences* which are not being observed by one individual/group to another. For example in our work in rural Wales, we found that the degree of intermixing or polarisation between new and established groups in rural Wales will depend in part on both the nature and intensity of the rural idyll imported by the in-migrant and the degree to which individuals within the locality cling to an identity of Welshness. Ideas relating to different scales of cultural construct and the potential of conflict when individuals or groups subscribe to different cultural norms, practices, symbols and visions do appear relevant in this kind of context, although this is an area that requires *very* careful research and interpretation if it is to go beyond the rather simple terms of explanation offered here.

IX

These four stories hardly scratch the surface of the task of enculturing political economy, yet as I have been involved in the reading, talking, interpreting and telling of these tales, my feeling remains that *both* the way in which opportunities are structured in rural areas *and* the ways in which rural lifestyles are experienced are together important in presenting an acceptable form of understanding of how change is being mapped out in the rural arena. Naturally, the use of terms such as 'acceptable' and 'understanding' is laden with unanswered questions: acceptable to whom and in what terms; understanding of what, the researched or the researcher? Yet these are questions which are common to all of us involved in these tasks so I find myself continuing to approve of the honesty demanded of authorship by the autoethnographic process. However, I am sure that however honest one intends to be in providing a personal context for writings and readings, it is probably impossible to convey with any accuracy the multiplicity of positions which characterise any particular set of interpretations. Whatever degree of self-reflexivity is conjured up in writing a piece like this, and more generally, there are constant implicit meanings associated with my maleness, whiteness, middle-classness, and so on which will be read by different readers into and from any story-telling I might construct. Indeed any attempts to practise what I called 'enculturing' broader political economy perspectives will be thus involved with these meanings, as well as crisscrossing with the meanings similarly implicit in the selves of different readers. Whilst accepting the inevitability of this complex crisscross of meanings, my immediate thought is that the individual self should not be elevated unduly in these processes. Any arrogance and pompousness of the first person singular needs to be avoided. Also, a strong appreciation needs to be cultivated that other lives, social practices and social processes exist and are reproduced regardless of the participation or non-participation of our selves, suggesting that the self is important in reading and writing these situations, but usually not to the practices and processes themselves.

In order to place into context this nagging interest in the conjunction of the structuring of opportunities and the experiencing of lifestyles, it is interesting to return to the three questions raised earlier: where one can speak from; to whom one speaks; and why one speaks at all. In honest answers to these questions lie important insights into the muddle and complexity of motive, purpose, practice and praxis in the life of an academic researcher. So, in the four stories outlined in the earlier part of this chapter, is it possible to say where I have been speaking from? My personal answer to this is that I have only been able to present a very partial location of these speaking positions. To the implicit social consciousnesses associated with gender, class, ethnicity, sexuality and so on, I have attempted to mix in the ingredients of specific living experiences, particularly influential people and texts, and more formal allegiances to spirtuality and political ideology. Inevitably even within these recognised influences I have omitted any number of important episodes, encounters and academic relations. How can we account for the specific influence of say conference papers, seminar presentations, journal articles, research team meetings and so on in the production of our stories? This perhaps begins to explain my own tendency to over-reference some of my writing, as well as the ethical stances on team-authorship within research projects which I hold dear.

Out of this *mélange* has consistently appeared an involvement with ideas of power and powerlessness. In each of the four stories told briefly above I have a nagging interest in the source, the nature, the impact and the reproduction of power within rural society as well as in the victims who are exploited or marginalised by the power of others and the powerlessness of themselves. This does prompt an interest in the political economic and managerial arrangement of available opportunities in rural areas, but it also begs questions about the notion of hegemonic social groups who are able to influence the social and cultural constructions of rurality and rural life which appear to set the stage for the *imagined* arena of rurality as a life-place and lifestyle. Reviewing the post-war history of rural studies it seems that the notion of power has been almost exclusively one which has political, economic or administrative connotations. Only relatively recently have the powers to construct and reconstruct socio-cultural expectations of space through language, through symbol, and through other forms of representation, been recognised as interesting and important elements in the arranging and experiencing of rural life. In my own case, the interest in socio-cultural geographies of the rural was pursued only during and even after the research project on new middle classes which brought the authors of this book together. So while I would be happy to suggest that I had an approach to the project all worked out before its beginning, the truth is that the approach that we started with was not the approach that we finished with, at least in part because our interviews with the people in the study villages led us to believe that a focus solely on the economic and political 'impacts' of new middle classes would lose sight of other influential components of rural lifestyle in these places. I suspect that this is often the case with the learning process which is involved in the practising of research.

Where I speak from, therefore, currently leads me to an interest in the different lines of power which crisscross rurality and rural life. Political-economic change continues to be important, but discursive and symbolic transformations are equally important in the understanding of imagined geographies of the rural. Some of the difficulties experienced by people living in rural areas will be political and economic, but others will spring from their marginalisation in socio-cultural terms. I therefore want to know about people's cultural competences in their chosen setting, about their strategies for rural living (whether explicit or implicit), and about the degree to which their lifestyle matches the cultural expectations raised by their imagined community and/or environment. In some senses this amounts to a focus on diversity and difference – recognising that given similar lifestyle opportunities people will experience their lives very differently. I have, however, already sought to explain my inability to accept fully the pluralism that has often accompanied post-modern moves in this direction. So within these stories of difference I find myself wanting to retain a strong politics of power and herein lie several muddling dilemmas, for it seems to me that broad politicised goals such as equality of opportunity sometimes require ideological principles of good/bad and right/wrong (both in the definition of which opportunities should be equalised, and in the revelation of unequal practices) yet at other times require a more pluralistic tolerance of different lifestyles, competences and strategies in rural locales. The ideology and the pluralism are linked, at least in my mind, because pluralistic tolerance will allow the exercise of social and cultural practices which result in power being expressed in that place, often serving to marginalise other people/groups in the process. At this stage, then, I would simply reiterate my own concern that important political values can be, and are being lost in some post-modern moves towards 'difference', and emphasise my own wish to recover a sense of power (including political power) whilst continuing to acknowledge difference in rural society. My other concern is that a focus on 'other', different groups and their neglected rural geographies will introduce a kind of 'fashion' effect into rural research whereby it will be 'right-on' to study particular facets of rural difference (gender, ethnicity, alternativeness, etc.) while other rural groups will be left out. By this I do not mean to diminish the importance of the subjects of rural 'right-on-ness' – indeed my own work is currently turning to some of these subjects – but rather to point out that less fashionable groups (Sarah Whatmore's example of these groups is people engaged in small farming business) are equally important to notions of other rural lifestyles.

The second question – to whom one speaks – is closely linked to these ideas relating to where one can speak from. Again taking the four stories of enculturing political economy as a starting point, a literal answer to the question of whom I was speaking to is fairly easy to explain. The work on 'structured coherence' with Mark Goodwin was written initially as a contribution to the Association of American Geographers conference in Miami and was subsequently rewritten into a paper for an academic journal. My story of 'new rural spaces for leisure' was a theme which had been bubbling away on the back-burner for a while, but was written formally when I was asked to contribute to a Festschrift in honour of Allan Patmore, who had been my Ph.D. examiner and whom I had got to know quite well subsequently. The book was on the theme of leisure and environment, and presented me with an unexpected opportunity to write under a banner headline – leisure – that I do not normally link myself with. The story about 'discursive transformations' with regard to poverty and rurality also began life as an AAG conference paper and has subsequently been rewritten as a journal article, and the work with Paul Milbourne on 'idylls' was published as a paper in the normal routine of publishing research findings. The literal answer then lies in the orthodox work practices of an academic researcher – books, articles, conferences – which can mean that I have been speaking to other academic researchers with similar interests in the rural, and who to an extent share the somewhat élite and restrictive languages and concepts used in academic story-telling.

If this were the only answer to the question of to whom one speaks, the following question (why one speaks at all) would be even more poignant. However, for most people in a similar job to mine there are in fact a wide range of other audiences to whom we are speaking. One obvious group who are spoken to (and indeed who speak back and thereby enrich and diversify the stories being told) are the university students at Bristol (and elsewhere) for whom some of these stories are lectures, seminars and supervisions. Another such group are the agencies and organisations who sponsor research, so over the last year, for example, I have given both verbal and written presentations to representatives from the Countryside Commission, Department of the Environment, Development Board for Rural Wales, Rowntree Foundation, Rural Development Commission, Welsh Development Agency and Welsh Office. I list these not as any form of self-glorifying curriculum vitae, but rather in an attempt further to answer the question of to whom one speaks. Neither do I suggest any particular influence with these agencies, because their reading of my presentations will again be subject – in institutional as well as individual form – to the histories and experiences of reflexivities discussed earlier as well as to the particular political and institutional constraints and profiles of the agencies concerned. To continue to list the actual audiences for these and other research stories over the last year, I have presented talks at a range of non-academic conferences (held, for example, by the National Farmers Union, the National Federation of Womens Institutes, Rural Community Councils, pressure groups, political groups etc.); given interviews to national newspapers and local television news programmes; acted as specialist adviser to a parliamentary committee; and conducted research interviews and meetings with numerous individuals and groups in rural areas including those with parish councils, school classes, Darby and Joan clubs, residents associations, and so on.

It is inevitable, given the diversity of these audiences, that stories will be told in very different languages, and sometimes that different stories will be told in the telling. Sometimes the nature of the story and the language of the telling are chosen by the story-teller. For example to speak at a conference organised by the NFU and NFWI and to suggest that many rural problems are seated in the hearts and minds of rural people such as myself and themselves, and further to illustrate the suggestion with reference to the discriminatory practices of many local rural people against ethnic minorities, alternative lifestyles and so on, represents a calculated personal wish to challenge some of the values held by some members of the audience (and indeed some of my own values). The fact that I was asked by the conference organisers to be 'controversial' helped enormously. Sometimes the 'audience' will insist on editorial control over how a story is constructed, and what is said. For example, recent experience of undertaking research for government departments has illustrated clearly to me that parts of a story provoke interest, other parts disinterest, and yet other parts an active form of censorship on the grounds of political acceptability or sensitivity. In this way, 'hard' numeric data is generally (but not exclusively) far more acceptable than 'soft' qualitative information, and 'policy-relevant' findings (relating say to houses, buses, medical services, training and enterprise) are more important to such audiences than 'policy-irrelevant' findings about social and cultural marginalisation and experiences of rural lifestyles. Similarly, notions of 'conflict', 'power' and 'need' can be very sensitive issues. The result in this and other similar cases, is the production of different versions of the story for different audiences – the formal report to research sponsors may be quite different from academic papers covering the same research.

To some extent we are talking here of the different discourses which are available to researchers. I am very comfortable with certain forms of academic discourse from which I have been receiving and to which I have been contributing over a number of years. The exploration of different experiences and imaginations of the rural will, however, lead to a wish to engage increasingly with lay discourses (see Halfacree, 1993), and indeed with the use of ethnographies there is a strong attempt to route academic discourses *through* lay discourses. This will involve a two-way translation process, interconnecting the voiced and otherwise represented experiences, attitudes and meanings of rural people with the processes of academic conceptualisation, as well as translating academic concepts into more widely understandable languages. In this way the story-teller receives from the audience as well as giving to it. This linking of lay and academic discourses is very relevant to the question of why one speaks at all. However, there appears to me to be real danger that in routeing academic discourses through lay discourses, these academic discourses will become incompatible with policy discourses. More traditional interconnections between academic research and policy-making agencies seem to have prospered precisely because academic discourses accepted many of the restrictions of research method, narrative and findings that were imposed by policy discourses. Now that lay discourses are beginning to be influential in the changing nature of academic discourse, these restrictions are being strongly challenged in some quarters with the result that lay/academic approaches can be incompatible with policy/academic approaches. This account inevitably simplifies the many important differences *within* these discursive categories, but it may nevertheless be the case that much more attention will have to be given to the issue of precisely how to include lay discourses without inducing discursive alienation from other groups that we might wish to talk to.

The question of 'why one speaks at all' is quite difficult to answer. In pragmatic terms, the 'why' is seated within a tradition of academic practice involving researching and writing articles and books which has been refocused by selectivity exercises from something fairly indistinct into more sharply defined sets of performance and output targets. Such a context does not, however, fully explain why a particular individual writes or tells stories. So part of my answer has to be that I speak because it is one of the conventions of my paid employment that I do so – the presentation of conference papers and the writing of journal papers and books are essential elements of the *modus operandi* of academic researchers – but other parts of my answer must point to the existence of a geographical imagination, a level of confidence, a vision for change and an energy to contribute which find expression in the academic context. Now these sound like grand claims, but I do not mean them to be so. I see it as part of any teacher's task to help their students to discover what is it that fires their imagination and to encourage them to make use of these imaginative and interpretative powers. Even though I am a teacher I am still a student, and so this does provide a personal motivation to speak and write. Alongside the imaginative quest, I also derive considerable enjoyment from the processes of researching and writing, and indeed from the interaction which follows with readers and listeners. Again to be a writer/speaker I believe you have also to be a reader/listener, and so these are essentially interconnected. Therefore, I speak because I enjoy doing so, and because I learn by doing so. In such circumstances, energy and confidence will abound in many people.

Perhaps it is the idea of a vision for change which will be most arguable. Many academic researchers will speak to inform, to influence others through that information and even to emancipate, although admitting the latter leaves one wide open to accusations of rather pompously perpetuating and reinforcing existing power networks. A recent debate in *Journal of Rural Studies* touches on this (Philo 1992, 1993; Murdoch and Pratt, 1993, 1994). Chris Philo (1992) suggests that rural geography should be made more open to the 'circumstances *and* to the voices of "other" peoples in "other" places: a new geography determined to overcome the neglect of "others" which has characterised much geographical endeavour to date' (p. 199).

Jon Murdoch and Andrew Pratt (1993) respond to this:

> In our view, what follows from this concern to 'give voice' are a set of issues which Philo does not really consider. . . . Simply 'giving voice' to 'others' by no means guarantees that we will uncover the relations which lead to marginalisation or neglect. This raises a whole clutch of issues relating to difference, space and power in relation to the 'rural'.
>
> (Murdoch and Pratt, 1993, p. 422)

Philo's reply (1993) takes up this challenge:

> Although I would add that in 'spirit' my own concern for injustice, exploitation and possible ways of improving conditions in the lives of rural 'others' is probably not so distant from the . . . one animating their

> approach to rural studies . . . I am still . . . unhappy about the assertive modernist impulse present in Bauman (and thus in Murdoch and Pratt) which proceeds with such certainty, which still puts faith in the a priori theoretical specification of how the world and its injustices operate, and which heroically assumes the duty of assessing from without the realities of 'other lives' against transcendental yardsticks of 'right'/'wrong' and 'good'/'bad' that may have little relevance for the peoples and places concerned.
>
> (Philo, 1993, p. 433)

Murdoch and Pratt's response (1994) is forthright, and endorses the idea of a vision for change:

> Should we not attempt to reveal the ways of the 'powerful', exploring the means by which they make and sustain their domination (perhaps in the hope that such knowledge could become a 'reservoir' to be drawn upon by oppositional actors)? Do we not also seek to influence the decisions of the 'powerful', such as policymakers, in the hope that they might be persuaded to produce more effective and just interventions in the world?
>
> (Murdoch and Pratt, 1994)

These debates are of the utmost importance in assessing the opportunities opened up for rural studies by post-modern and post-structural approaches. Essentially, Philo's post-modern rural geographies would stop short of 'transcendental' yardsticks against which to challenge policy-makers to change, preferring a more pluralistic way of thinking which does not as of right elevate particular visions of what is right or good. Murdoch and Pratt's post-modern rural studies are to them the most exciting approach with which to explore and understand the use and misuse of power in rural areas.

For me the Christian and political impulses are too strong to accept the pluralist position. I like the idea of providing a channel through which other voices can be heard and I like the ways in which qualitative methodologies are pursuing this aim. Having heard these voices, however, I am not personally content to grant them equal credence in my own interpretation of what is and what could be. I therefore like Murdoch and Pratt's desire to go beyond the 'giving voice' to the exploration of the powerful and the influencing of policy-makers, and for me this is part of the answer to why I speak. Murdoch and Pratt's notion of 'more effective and just interventions', however, does not sit easily in the post-modern idiom (as Philo points out), a disjuncture which reinforces my apprehensiveness towards that label. Nevertheless I for one will be most interested to read other accounts of these issues, and expect that rural studies will become more thoughtful, and perhaps even more influential in the process. Perhaps that is, after all, why one speaks.

REFERENCES

Ang, I. (1989) Beyond self-reflexivity, *Journal of Communication Inquiry*, Vol. 13, no. 2, pp. 27–9.

Atkinson, P. (1990) *The Ethnographic Imagination: Textual Constructions of Reality*, Routledge, London.

Barnes, T. and Duncan, J. (eds.) (1992) *Writing Worlds: Discourse, Text and Metaphor in the Representation of Landscape*, Routledge, London.
Baudrillard (1983a) *Simulations*, Semiotext, New York.
Baudrillard (1983b) *In the Shadow of the Silent Majority*, Semiotext, New York.
Best (1989) The commodification of reality and the reality of commodification: Jean Baudrillard and post-modernism, *Current Perspectives in Social Theory*, Vol. 19, pp. 23–51.
Bristow, J. (1991) Life stories: Carolyn Steedman's history writing, *New Formations*, Vol. 13, pp. 113–31.
Christian, B. (1989) But what do we think we're doing anyway: the state of black feminist criticism(s) or my version of a little bit of history, in C. Wall (ed.) *Changing our own Words: Essays on Criticism Theory and Writing by Black Women*, Rutgers University Press, New Brunswick.
Clifford, J. (1986) On ethnographic allegory, in J. Clifford and G. Marcus (eds.) *Writing Culture*, University of California Press, Berkeley.
Cloke, P. (ed.) (1987) *Rural Planning: Policy into Action*, Harper & Row, London.
Cloke, P. (1989) Rural geography and political economy, in R. Peet and N. Thrift (eds.) *New Models in Geography*, Unwin Hyman, London.
Cloke, P. (1992) The countryside, in P. Cloke (ed.) *Policy and Change in Thatcher's Britain*, Pergamon, Oxford.
Cloke, P. (1993) The countryside as commodity: new rural spaces for leisure, in S. Glyptis (ed.) *Leisure and the Environment*, Belhaven, London.
Cloke, P. (1994) Rural poverty and the welfare state: a discursive transformation in Britain and the USA, *Environment and Planning A*, forthcoming.
Cloke, P. and Goodwin, M. (1992) Conceptualising social change: from post-Fordism to rural structured coherence, *Transactions IBG*, NS 17, pp. 321–36.
Cloke, P. and Goodwin, M. (1993) The changing function and position of rural areas in Europe, *Nederlandse Geografische Studies*, Vol. 153, pp. 19–36.
Cloke, P. and Little, J. (1990) *The Rural State*? Oxford University Press.
Cloke, P. and Milbourne, P. (1992) Deprivation and lifestyles in rural Wales: II Rurality and the cultural dimension, *Journal of Rural Studies*, Vol. 8, pp. 359–71.
Cloke, P. and Thrift, N. (1990) Class and change in rural Britain, in T. Marsden, P. Lowe and S. Whatmore (eds.) *Rural Restructuring*, David Fulton, London.
Cohen, A. (1983) *Belonging*, Manchester University Press.
Cooke, P. (1985) Class practices as regional markers: a contribution to labour geography, in D. Gregory and J. Urry (eds.) *Social Relations and Spatial Structures*, Macmillan, London.
Dean, H. (1991) In search of the underclass, in P. Brown and R. Scase (eds.) *Poor Work*, Open University Press, Buckingham.
Fiske, J. (1990) Ethnosemiotics: some personal and theoretical reflections, *Cultural Studies*, Vol. 4, pp. 85–99.
Foley, W. (1974) *A Child in the Forest*, BBC Publications, London.

Fuss, D. (1989) *Essentially Speaking: Feminism, Nature and Difference*, Routledge, London.
Geertz, C. (1988) *Works and Lives: The Authropologist as Author*, Stanford University Press.
Gilbert, A. (1988) The new regional geography in English and French-speaking countries, *Progress in Human Geography*, Vol. 12, pp. 208–28.
Goodwin, M., Duncan, S. and Halford, S. (1993) Regulation theory, the local state, and the transition of urban politics, *Society and Space*, Vol. 11, pp. 69–88.
Gramsci, A. (1971) *Selections from the Prison Notebooks*, Lawrence & Wishart, London.
Grossberg, L. (1989) On the road with three ethnographers, *Journal of Communication Inquiry*, Vol. 13, no. 2, pp. 23–6.
Halfacree, K. (1993) Locality and social representation: space, discourse and alternative definitions of the rural, *Journal of Rural Studies*, Vol. 9, pp. 1–15.
Hall, S. (1988) The toad in the garden: Thatcherism among the theorists, in C. Nelson and L. Grossberg (eds.) *Marxism and the Interpretation of Culture*, Macmillan, London.
Handler, J. and Hasenfeld, Y. (1991) *The Moral Construction of Poverty*, Sage, Newbury Park, CA.
Harvey, D. (1985) *The Urbanisation of Capital*, Blackwell, Oxford.
Hoggart, K. (1990) Let's do away with rural, *Journal of Rural Studies*, Vol. 6, pp. 245–57.
James, S. (1991) *The Urban–Rural Myth – or Reality?* Department of Geography, University of Reading, Geographical Papers No. 107.
Jessop, B. (1990) Regulation theories in retrospect and prospect, *Economy and Society*, Vol. 19, pp. 153–216.
Keating, M. (1988) *State and Regional Nationalism*, Harvester Wheatsheaf, London.
Knox, P. and Cottam, B. (1981) Rural deprivation in Scotland: a preliminary assessment, *Tijdschrift voor Economische en Sociale Geografie*, Vol. 72, pp. 162–75.
Layder, D. (1993) *New Strategies in Social Research*, Polity Press, Oxford.
Mack, J. and Lansley, S. (1985) *Poor Britain*, Allen & Unwin, London.
Massey, D. (1988) A new class of geography, *Marxism Today*, May, pp. 12–17.
Mingay, G. (ed.) (1989a) *The Rural Idyll*, Routledge, London.
Mingay, G. (ed.) (1989b) *The Unquiet Countryside*, Routledge, London.
Mingay, G. (ed.) (1989c) *The Vanishing Countryman*, Routledge, London.
Mormont, M. (1990) What is rural? or, How to be rural: towards a sociology of the rural, in T. Marsden, P. Lowe and S. Whatmore (eds.) *Rural Restructuring*, David Fulton, London.
Murdoch, J. and Pratt, A. (1993) Rural studies: modernism, postmodernism and the 'post-rural', *Journal of Rural Studies*, Vol. 9, pp. 411–28.
Murdoch, J. and Pratt, A. (1994) Rural studies of power and the power of rural studies: a reply to Philo, *Journal of Rural Studies*, Vol. 10, pp. 83–8.
Nairn, T. (1977) *The Break-up of Britain*, New Left Books, London.

Newby, H. (1987a) *Countryside Life: A Social History of Rural England*, Wiedenfeld & Nicholson, London.

Newby, H. (1987b) *The Countryside in Question*, Hutchinson, London.

Oppenheim, C. (1990) *Poverty: The Facts*, Child Poverty Action Group, London.

Peet, R. and Thrift, N. (1989) Political economy and human geography, in R. Peet and N. Thrift (eds.) *New Models in Geography*, Unwin Hyman, London.

Philo, C. (1992) Neglected rural geographies: a review, *Journal of Rural Studies*, Vol. 8, pp. 193–207.

Philo, C. (1993) Postmodern rural geography? A reply to Murdoch and Pratt, *Journal of Rural Studies*, Vol. 9. pp. 429–36.

Probyn, E. (1993) *Sexing the Self: Gendered Positions in Cultural Studies*, Routledge, London.

Radway, J. (1988) Reception study: ethnography and the problems of dispersed audiences and nomadic subjects, *Cultural Studies*, Vol. 2, no. 3, pp. 359–67.

Radway, J. (1989) Ethnography among elites: comparing discourses of power, *Journal of Communication Inquiry*, Vol. 13, no. 2, pp. 3–11.

Said, E. (1979) *Orientalism*, Vintage Books, New York.

Short, B. (ed.) (1992) *The English Rural Community*, Cambridge University Press.

Short, J. (1991) *Imagined Country: Society, Culture and Environment*, Routledge, London.

Spivak, G. (1989) Who claims alterity?, in B. Kruger and P. Mariani (eds.) *Remaking History*, Bay Press, Seattle.

Stanley, L. (ed.) (1990) *Feminist Praxis: Research, Theory and Epistemology in Feminist Sociology*, Routledge, London.

Steedman, C. (1986) *Landscape for a Good Woman*, Virago Press, London.

Williams, R. (1973) *The Country and the City*, Chatto & Windus, London.

Woodward, R. (1992) *Deprivation and Marginalisation in Rural Areas*, Rural Lifestyles Project Working Paper 1. St David's University College, Lampeter.

Chapter 5

Inhuman Geographies: Landscapes of Speed, Light and Power

Nigel Thrift

> We live us as others allow us to live, creating meeting places as we go along. Such places may not be monumental, they may be like nothing more than encounters, spatial events that leave behind them less litter than a campsite, yet they can form the basis of a social fabric, one that does not suppress the contingency of its community but makes its migratory haphazardness the material out of which it weaves its identity.
>
> (Carter, 1992, p. 8)

> Between things does not designate a localisable relation going from one to the other and reciprocally, but a perpendicular direction, a transversal movement carrying away the one and the other, a stream without beginning or end, gnawing away at its two banks and picking up speed in the middle.
>
> (Deleuze and Guattari, 1983, p. 58)

INTRODUCTION

This chapter is, I am afraid, yet another attempt to understand the nature of contemporary western societies. My thoughts on 'modernity' – a word which, by the way, I dislike[1] – have been crystallised by consideration of a commonplace, even banal, image; an urban landscape at night through which runs a river of headlight. This frozen image of mobility summarises a number of themes that I want to take up in this paper – speed, light, and power. At the same time, it also signifies a shift in the realm of human experience, a shift which we still find difficult to imagine, towards what Haraway (1985, 1991a, 1991b, 1992) has called a 'cyborg culture', a culture of foregrounded codes

[1] I dislike the word 'modernity' for two reasons. First, because it implies that there is some superior form of experience (Sayer, 1991, p. 63–72). Second, because it perpetuates the idea that there are 'beginnings, enlightenments, and endings: the world has always been in the middle of things' (Haraway, 1992a, p. 304). See also Latour (1993).

and redundancies in which the boundaries between people and 'machines'[2] have started to break down.

What I want to suggest is that we have now reached a point where western cultures have become increasingly self-referential in the sense that, over a number of generations, sources and horizons of meaning have developed and become generalised, sedimented and then mutated, which are based in hybrid images of machine and organism, especially images based on speed, light, and power. As a result, we now live in an almost/not quite world – a world of almost/not quite subjects; almost/not quite selves; almost/not quite spaces and; almost/not quite times – which has become one of the chief concerns of contemporary experience and social theory. This chapter is an attempt to expand on and document this cultural hypothesis.

Accordingly, the chapter is in five main parts. In the first part I want to resurrect the concept of a structure of feeling. I will utilise this concept in the remainder of the paper as a means of understanding the almost/not quite status of an intellectual project which I have christened 'mobility'. This project is the attempt to describe, and begin to theorise, new orders of experience constructed out of 'machinic' sources and horizons of meaning (Kenner, 1987). In this part of the paper I will also attempt to provide some notion of the contemporary social and cultural conditions of academe out of which this structure of feeling has arisen. However, much of this structure of feeling is clearly the result of much wider-ranging historical changes in discourse networks. Therefore, in the second part of the paper, I will document some of these changes, concentrating in particular on the machinic complexes of speed, light and power. This historical excursion allows me to then to go on to outline the key elements of an almost/not quite ontology which is gradually gathering momentum around the key trope of mobility. In the fourth part of the chapter, I will try to show some of the ways in which this structure of feeling can be related to changing research agendas in the social sciences and humanities, concentrating in particular on geographical work. Finally, I will provide a set of conclusions.

Some readers will no doubt enquire as to what such a chapter is doing in a book on countryside. I offer no apologies. It is not just that the countryside is nearly always connected with a degree of artifice. It is not just that countryside has multiple meanings arrayed across different social groups and cultures. It is not just that our ways of seeing countryside have been technologised. These are all, essentially, interpretive concerns. Rather, it is that the ground of what we count as experience has shifted and, as a result, interpretation is no longer an entirely adequate communicational technique. Interpretation works with constants, what we now light on are variables. In other words, both the texture of 'reality' and its recording threshold have changed and 'Countryside' has

[2] I use machinery and machinic in the widest possible sense here and in the rest of the paper. The terms are not meant to imply a machinistic model of reality. Rather, they signify the existence of a set of machineries of desire (see Bogue, 1989, pp. 81–91) and they signal that nowadays 'our machines are disturbingly lively, and we ourselves frighteningly inert' (Haraway, 1991a, p. 144).

become mixed up in this drama. The discourse network exceeds the word, so to speak. And it is this discourse network which I want to address.

1. AN EMERGENT STRUCTURE OF FEELING

How is it possible to describe an almost/not quite intellectual project? That requires an almost/not quite concept. Such a concept is available in the form of Raymond Williams' (1954, 1977, 1979, 1981) notion of a 'structure of feeling'. Williams' conception is, of course, notoriously elusive. On one level, the term is simply intended to signify 'the culture of a period' (Williams, 1979, p. 48). But on another level it is an attempt to get at something more elusive; 'the *living* result of all the elements in the general organisation' (Williams, 1979, p. 48, my emphasis). Yet it is this very elusiveness which is particularly appropriate for my purposes, for at least five reasons. First of all, the idea of structure of feeling is intended to signal a *process* that is 'at the very edge of semantic availability' (Williams, 1977, p. 134), echoing a concern that nowadays is more likely to be addressed by terms like liminality and differend. Second, the notion fits well with the reading of current intellectual tendencies that I want to make. Williams wanted to stress the continual mobility of modern cultural processes; always 'present', 'moving', 'active', 'formative', 'in process', 'in solution'. Williams' interest was quite clearly to avoid reducing many aspects of the cultural to a fixed 'form'. Rather, there are traces, and traces of traces.[3] Third, Williams stresses over and over again that structures of feeling are never reducible to the institutional or formal. 'What really changes is something quite general, over a wide range, and the description that often fits the change best is the literary term "style"' (Williams, 1977, p. 131). Thus Williams envisaged structures of feeling as *cultural hypotheses* 'that do not have to await definition, classification, or rationalisation before they exact palpable pressures and set effective limits on experience and action' (Williams, 1977, p. 132). Fourth, the idea of structure of feeling is intended to signal a commitment to *experience*, to the fact that experience is something which is not reducible. It is what we live through. It is 'the movement of bone, of body, of breath, of imagination, of muscle, and the conviction of sheer stubborness that there are other possibilities' (Probyn, 1993, p. 172). It is that 'element for which there is no external counterpart' (Williams, 1954, p. 21). Thus, structure of feeling necessarily involves calling on emotion, bodily practices, the physical character of places, all those differentiated and differentiating elements of comportment which are crucial to a culture, may well be bound up with reading and writing, yet are so difficult to read and write (Probyn, 1993; see also Scott, 1992). Finally, the notion itself can be seen, in some senses, at least, as not merely methodological but as actually reflecting a specific historical period in which almost/not quite effects become more

[3] It becomes clear, I think, that in his latest work Williams may have had rather more in common with Derrida than is widely acknowledged.

palpable. In other words, the notion of structure of feeling can also describe the practical penumbra of some current intellectual developments. It describes a kind of shadow world which is now coming out into the light.

Where has this structure of feeling which I call mobility actually come from? What factors underlie such a redistribution of intellectual interest? To answer these questions requires a consideration of both the changing social field of academe and the nature of wider changes in the world with which this field must interact.

Certainly, over the last twenty years or so, there have been a number of important changes in what Bourdieu (1988, 1990, 1991, 1992) calls the social field of academe, which have led to a greater emphasis on culture and cultural theory and, in turn, have prepared the ground for the new structure of feeling. To follow Williams (1977) again, these changes have been a mixture of the institutional (most especially, major changes in higher education systems in a number of countries), the formal (in particular, the rise of poststructuralism and the associated transformation of academic heretics into establishment heresiarchs (Bourdieu, 1988; Lamont, 1987)) and the social (especially the gradual, uneven but still marked shift in the balance of intellectual power in academe based on the interests of new generations, new social groups, and new class fractions.)[4] These different but cross-cutting changes in the social field have certainly all contributed to the new structure of feeling in various ways. The importance of institutions can be seen in the way that, in Britain, for example, it was the former polytechnics, with their clearer interests in technology, that also took up cultural studies most keenly. The importance of forms can be seen in the way that many poststructuralist authors have used machinic means of metaphorisation. The importance of new generations and new social groups can be seen in the way that matters of gender, sexuality and race have been taken up and have led to much greater attention being given to borders, transgression, third cultures and other motifs of the new structure of feeling, while the importance of new class fractions can be seen in the role of academics in the rise of a new petit bourgeoisie (Bourdieu, 1984) with its base in the cultural industries and its stress on the construction and manipulation of images and related acts of consumer living.

Quite clearly the new structure of feeling has been influenced by these changes. However, equally clearly, it cannot 'without loss, be reduced to belief systems, institutions, or explicit general relationships' (Williams, 1977, p. 133). I want to claim that to get at the 'generative immediacy' of the new structure of feeling any analysis has to go much further, to how such social changes are interpreted, in this case as a remetaphorisation based in machinic complexes which I have termed, speed, light, and power, which goes beyond the old organic and technological metaphors (Haraway, 1992a, p. 21). In other words, it is important to get at something we might loosely call 'a change

[4] One might also point to the increased international mobility of academics as another important factor!

of style', which also turns out to be a change in content.[5] That is the main purpose of this chapter.

By taking such a view, it will hopefully become apparent that the new structure of feeling is an attempt to articulate and be articulated by a cyborg culture (Haraway, 1985, 1989, 1991a, 1991b, 1992a, 1992b). Following Haraway, such a culture can be described as a hybrid of machine and organism, 'a condensed image of both imagination and material reality, the two joined centres structuring any possibility of historical transformation' (Haraway, 1989, p. 191). In other words, it is a re-recognition of the difficulties associated with projects that divide the world into the human and the nonhuman, that insist that objects are 'shapeless receptacles of social categories' (Latour, 1993, p. 55) without agency, and that deny the multiplication of what Latour (1993) calls 'quasi-objects' and Haraway calls 'actants', collective ensembles 'which act, (have) will, meaning and even speech' (Latour, 1993, p. 136). It is to point to the increasing difficulty of situating 'the human', to restore 'the share of things' (Latour, 1993, p. 136) and to recognise the increasing importance of 'in-between' life forms (Michaux, 1992).

That the structure of feeling I have called 'mobility' has emerged at this particular time I take to be no accident. In particular, I take it that it has been forced as a result of three closely connected imaginative adjustments. The first of these stems from the fact that we live in an increasingly artificial, or more accurately, manufactured environment. The transformation to this kind of environment has taken many hundreds of years and is of immense cultural significance. Williams (1990) compares it, in its scope and effect, to the transition from hunter-gatherer to agricultural civilisations which also took many hundreds of years and which led to profound moral, political, and spiritual upheavals as the one world was replaced by another.

> It is not only imaginable but probable that humanity's decision to unbind itself from the soil – not to return to a nomadic existence but to bind itself instead to a predominantly technological environment – has provoked a singularly profound spiritual crisis. We are now embarked upon another period of cultural mourning and upheaval, as we look back to a way of life that is ebbing away.
>
> (Williams, 1990, p. 2)

Or, as Jameson (1991, p. 35) puts it, 'the other of our society is in (a) sense no longer Nature at all, as it was in precapitalist societies, but something else which we must now identify'.

The second imaginative adjustment is to the scientific advances which have enabled this transition to take place. The structure of feeling called mobility emerges at a time of a kind of disillusionment with the science which has produced a manufactured environment. This disillusionment takes a number of forms. First, there are the various forms of 'life politics' (Giddens, 1991)

[5] The term style is used here in the same sense as it is used by Wood (1990).

which arise out of a greater level of reflexivity concerning scientific knowledge. Second, there is the anthropological importance of science.

> Threats from civilization are bringing about a new kind of 'shadow kingdom', comparable to the realm of the gods and demons in antiquity, which is hidden behind the visible world and threatens human life on this Earth. People no longer correspond today with spirits residing in things, but find themselves exposed to 'radiation', ingest 'toxic levels' and are pursued into their very dreams by the anxieties of a nuclear holocaust. The place of the anthropomorphic interpretation of nature and the environment has been taken by the modern risk consciousness of civilisation with its imperceptible and yet omnipresent latent causality.
> (Beck, 1992, p. 72)

Third, there is a more general solidarity with other living things which is, at the same time, a renegotiation of the body and the self. Thus, awareness of the body increases; 'once thought to be the locus of the soul, then the centre of dark perverse needs, the body has become fully available to be "worked upon"' (Giddens, 1991, p. 218). Relatedly, the self becomes linked to 'nature'. 'Where trees are cut down and animal species destroyed, people feel victimised themselves in a certain sense. The threats to life in the development of civilisation touch commonalities of the experience of organic life that connect the human vital necessities to those of plants and animals' (Beck, 1992, p. 74). Finally, partly because much of what we regard as 'nature' can no longer survive without human intervention (Strathern, 1992a), society is no longer understood as separate from nature; 'the "end of nature" means that the natural world has become in large part a "created environment", consisting of humanly structured systems whose motive, power and dynamics derive from socially organised knowledge–claims rather than from influences exogenous to human activity' (Giddens, 1991, p. 144). At the same time, the disillusionment with science also arises, in part, from an accommodation with it; science and its doings are no longer perceived as something 'new', 'modern' or 'progressive' but as simply a part of the fabric of everyday life, and as therefore subject to the same moral judgements (Beck, 1992).

The third imaginative adjustment stems from a re-view of machines and machinic complexes. Machines are no longer conceived of in the same way. They are no longer viewed as either a threat or a salvation but as troublesome companions (Bijker, Hughes and Pinch, 1989; Bijker and Law, 1991). The view, running from Babbage and Ure through to Fritz Lang's *Metropolis*, of a world of 'vast automatons' in which the machine is 'a creation destined to restore order amongst the industrious classes' (Ure, 1835, cited in Naylor, 1992, p. 228) has been replaced by a view of machinery that is more ambiguous. Increasingly, machines are no longer viewed as a separate realm but, to use Haraway's term, they are seen as functioning parts of 'actants' – agencies in which the 'actors' are not all 'us':[6] 'this is the barely admissible recognition

[6] 'Non humans are not "actors" in the human sense, but they are part of the functional collective that makes up an actant' (Haraway, 1992a, p. 331).

of the odd sorts of agents and actors which/whom we must now admit to the narrative of collective life' (Haraway, 1992a, p. 297). Such a recognition clearly requires a number of further imaginative compromises. First, there is the nature of what it is to be human. Thus machines can no longer be automatically regarded as without human characteristics. Indeed perhaps we now need to talk of a new category of people/machines – the inhuman (Lyotard, 1992). Second, there is the nature of the subject. Subjects can no longer be seen as bounded by bodies. Or to put it another way, subjects can no longer be seen as fixed nodes. They are part of human–machine networks of social connectedness that change what it means to be human (Kittler, 1990). '*Ecce homo*: delegated, mediated, distributed, mandated, uttered' (Latour, 1993, p. 138). Third, there is the nature of action: 'action is not so much an ontological as a semiotic problem' (Haraway, 1992a, p. 331, fn. 11). In other words, what counts as an actor?.[7]

This, then, is something of what is meant by a change of style! In the next section of this paper I will begin to articulate this structure of feeling by referring to the history of three of the main machinic complexes – speed, light and power[8] – which, as we shall see, very soon begin to collapse into each other. Speed, light, power; mobility.

2. SPEED, LIGHT, POWER; MOBILITY

A machinic complex is here taken to be a developing bundle of institutions and technologies, understood as systems of knowledge–discipline-perception based in a 'parliament of things': embodied subjects, machines, texts and metaphors, and the like (Haraway, 1992a, 1992b, 1992c), which undergo periodic modernisation, redirection and redefinition. In this section, I will provide a brief history of the three different machinic complexes of speed, light, and power. The history of these three complexes has been a highly uneven one, both between and within different western countries. To give some sense of this unevenness, I therefore make particular reference to the case of Britain, a country which has been less mobile than some (for example, the United States with its level of 1.3 people per automobile which Baudrillard (1988, p. 27), with, for once, understatement has termed 'pure circulation') but more mobile than others. The section starts out by referring to the nineteenth century when it is still possible to treat these machinic complexes as separate and separated. However, in the latter parts of the section, I will treat these three complexes as one, a recognition of the way in which they have now come together in a kind of social-technological jouissance.

I am well aware of the dangers in this kind of account of a latent or explicit technological determinism in which 'an independent dynamic of mechanical

[7] As will become clear, this viewpoint differs from that of many authors who believe that 'species' dividing lines are still clear cut (e.g. Benton, 1993).

[8] I am well aware that there are other complexes that could also have been fixed on, and especially those to do with the machinic complex of sound (Attali, 1986) and touch but the addition of these complexes would have made the chapter even more unwieldy.

invention, modification, and perfection imposes itself on to a social field transforming it from outside' (Crary, 1990, p. 8). However, at the same time, it is especially important not to produce a sociological or cultural determinism in which socio-cultural forces produce objects. Part of the point of this chapter is that, in Latour's (1987) terms, the 'technogram' is the other side of the 'sociogram' and 'every piece of information you obtain on one system is also information on the other'. Or, as Deleuze and Guattari (1988, p. 90) put it: 'a society is defined by its amalgamations, not by its tools . . . tools exist only in relation to the intermingling they make possible or that make them possible'. In other words, we must see technologies as parts of networks, made up of actors (of whatever kind), and these quasi-objects are greater than the sum of their parts. But none of this is to suggest that these assemblages are neutral. They embody technical codes[9] (which might be seen as general rules for connecting technograms and sociograms) which are institutionalised in various ways, and these institutions have margins of manouvre associated with them which can allow the institutions and even codes to be redefined (Feenberg, 1991).

2.1 The nineteenth century

2.1.1 Speed

I hardly need to reprise a history of the machinic complex of speed that results from the application of new technologies of transport and communication from the end of the eighteenth century onwards (Thrift, 1990). I therefore provide here only the briefest sketch.

In the nineteenth century the technology of speed broke through the limits of walking and the horse into a period of progressively accelerating transport network technology – the stage coach and the horse-drawn tram, the railway and the electric train, and the bicycle. Thus in Britain by 1820 it was often quicker to travel by stage coach than on horseback. By 1830 movement between the major towns was some four or five times faster than in 1750. This increase in the speed of the stage coach was paralleled by an increase in both frequency of operation and an increase in the number of destinations. The subsequent growth of the railway network made for even more dramatic leaps in speed, frequency, and access. It is no surprise that 'the annihalation of space by time' was a favourite meditation for the Victorian writer.[10] The effects were all the more arresting because they came to be experienced by so many people. By 1870, 336.5 million journeys were made by rail, the vast bulk of them by third-class passengers. In the growing cities a parallel process of democratisation was taking place measured out by the advent of the horse-drawn tram, the underground, and, latterly, electric tramways. But perhaps the most dra-

[9] Feenberg (1991) specifically links such codes to Foucault's notion of regimes of truth.

[10] It is only a surprise that so many commentators seem to believe that Marx invented the term.

matic change in travel was the invention of the bicycle. By 1855 there were already 400,000 cyclists in Britain and the 1890s saw the peak of this simple machine's popularity. The bicycle, which started as a piece of fun for young swells, foreshadowed the automobile in providing immediate, democratic access to speed (Kern, 1983).

The nineteenth century also saw the beginnings of new networks of communication which began to displace face-to-face communication, especially a rapid and efficient mail service and, latterly, the telegraph and mass-circulation newspapers. In Britain, the mail service expanded massively in the nineteenth century, as did new communication innovations like the postcard, valentine cards, Christmas cards, and so on. By 1890, the Post Office was carrying 1,706 million letters a year. The telegraph was first used in 1839. By 1863 nearly 22,000 miles of line had been set up, transmitting over six million messages a year, from 3,381 points. However, it was not until the last quarter of the nineteenth century that the telegraph became an institution of communication genuinely used by the mass of the population. However, neither of these means of communication was instantaneous. The mail service still depended upon the velocity of a set of different means of transport while the telegraph service had to be actually reached (usually in a post office) before it could be used.

The exact social and cultural effects of this 'great acceleration' have certainly been disputed. But, amongst the changes attendant on a world of traffic flowing through multiple networks at least four might be counted as being significant. The first of these was a change in the consciousness of time and space. For example, so far as time consciousness was concerned, it is clear that the population began to pay more attention to smaller distinctions in time (Thrift, 1990). Thus, in the last decade of the nineteenth century watches became more popular (Kern, 1983). Again, it is clear that a sense of an enlarged simultaneous presence became more common, especially as a result of the telegraph (Kern, 1983; Briggs, 1989). This was not just a temporal but also a spatial sense. In one sense space had been shrunk by the new simultaneity engendered by the shrinkage of travel and communication times, and by new social practices like travelling to work and tourism. In another sense it was much enlarged. *The Times* of 1858 (quoted in Briggs, 1989, p. 29) wrote proudly of 'the vast enlargement . . . given to the sphere of human activity' by the telegraph and the press that now fed upon its pulses. It is also possible to identify elements of a change in the perception of landscape, one that views landscape as something seen from within a moving platform. Schivelsbuch (1986) has called this new appreciation a 'panoramic perception' (see also Kern, 1983; Tichi, 1987) in which the world is presented as a passing, momentary spectacle to be glimpsed and consumed.[11] A second change was the diffusion of this general speed-up into the texts of the period, whether in the form of enthusiastic paeans to machinery (see Naylor, 1992) or in the form of

[11] However, it is important to note that this consciousness of space was not all one-way. The new large cities were often depicted as of endless extent, rather than as shrunken spaces.

counter-laments for slower, less mechanical times gone by. A third change was in the nature of subjectivity. In particular, it is possible to note an increasing sense of the body as an anonymised parcel of flesh shunted from place to place, just like other goods. Each body passively avoided others, yet was still linked in to events, at a distance, a situation typified by Victorian vignettes of the railway passenger in greater communion with the newspaper than companions (Schivelsbuch, 1986). 'Each individual paper, a replica of hundreds of thousands of others, served as a private opening to a world identical to that of one's companion on a street car, a companion likely to remain as distant, remote and strange as the day's news came to seem familiar, personal, real' (Trachtenberg, 1982, p. 125). A final change was in the metaphors that were becoming dominant, especially in bourgeois circles. Chief amongst these were metaphors of 'circulation' (usually via a semantic mixture of the bodily and the mechanical) and 'progress'. Thus speed; expansion; abundance (Buck Morss, 1990) (connections usually made via a semantic operation in which spatial movement via a particular technology like the railway was wedded to historical movement). 'The formula is as simple as can be: whatever was part of circulation was regarded as healthy, progressive, constructive; all that was detached from circulation, on the other hand, appeared diseased, medieval, subversive, threatening' (Schivelsbuch, 1986, p. 195). Schivelsbuch is able to tie these metaphors of circulation and progress together through this formula:

> The notion that communication, exchange and motion bring to humanity enlightenment and progress, and that isolation and disconnection are the obstacles to be overcome on this course, is as old as the modern age. The bourgeois cultural development of the last three centuries can be seen as closely connected with the actual development of traffic. In retrospect, it is easy to see what significance the experience of space and time had for bourgeois education when one considers the Grand Tour, which was an essential part of that education before the industrialisation of travel. The world was experienced in its original spatio-temporality. The travelling subject experienced localities in their spatial individuality. His education consisted of his assimilation of the spatial individuality of the places visited, by means of an effort that was both physical and intellectual. The eighteenth century travel novel became the *bildungsroman* (novel of education) of the early nineteenth century. The motion of travel, that physical and intellectual effort in space and time, dominated both.
>
> The railroad, the destroyer of experiential space and time, thus also destroyed the experience of the Grand Tour. Henceforth, the localities were no longer spatially individual or autonomous: they were points in the circulation of traffic that made them accessible. As we have seen, that traffic was the physical manifestation of the circulation of goods. From that time on, the places visited by the traveller became increasingly similar to the commodities that were part of the same circulation system. For the twentieth century tourist, the world has become one huge department store of countrysides and cities.
>
> (Schivelsbuch, 1986, p. 197)

2.1.2 Light

Another important part of the history of the machinic complex of mobility has been the history of the machinic complex of artificial light. In effect, this history dates from the end of the eighteenth century, when, for the first time, a technology that had not significantly altered for several hundred years, began to change. Before this time, artificial light had been in short supply. To an extent, the work day had been emancipated from dependence on daylight by candles and oil lamps, but most households used artificial light only very sparingly (Schivelsbuch, 1988).

From the end of the eighteenth century through the early nineteenth centry a whole series of inventions began to change this situation. The Argand oil-burning light (1783) was the first such invention. It was soon followed by the gaslight which was made possible by the invention of systems to produce gas from coal circa 1800. These were first used in factories and then, with the setting up of central gas supplies in cities (the first gasworks was set up in Britain in London in 1814), they spread to the household. The networks of gas mains prefigured railway tracks and electrical networks. By 1822 there were already 200 miles of gas main in London, and 53 English cities in all had gas companies. By 1840 there was 'scarcely a town of any importance' that was not lit with gas (Clegg, 1841, cited in Robson, 1973, p. 178). The gas companies fuelled gas lamps in the household and in the street (although, as Robson (1973) shows, adoption was very uneven).

To some extent running in parallel with the development of gas lighting was the development of brighter and more spectacular electrical lighting. The arc light was invented in 1800 and was used in specific situations over long periods of time but was not put into general use in factories, shops and similar sites until the 1870s and 1880s. It was the invention of the electrical bulb in 1879 which heralded the widespread electrification of light and the decisive break between light and fire. The first central electricity generating stations became operational in 1882 in New York and London. However, although London was amongst the first to have a fully functioning electric light system (around Holborn) subsequently, like Britain as a whole, it lagged in the adoption of electric light, partly because of economic conditions, partly because of legislation – which prevented the growth of electricity monopolies – and partly because of opposition from gas interests. Thus, it was not until the 1920s that electricity, and electrical light, was widely adopted in Britain (Hughes, 1983; Byatt, 1979). In contrast, by 1903 New York had 17,000 electric street lamps, 'while electric interior lighting and exterior displays had become "essential to competition" for downtown theatres, restaurants, hotels and department stores' (Nasaw, 1992, p. 274).

This history of an ever-expanding landscape of light which we now take so much for granted cannot be ignored. As a machinic complex it is particularly important because of five changes. The first of these is that it signifies the progressive colonisation of the night (Melbin, 1987). Night is no longer regarded as a period of general inactivity. Communities start to become incessant; there is 'a blurring of the division of day and night' (Kern, 1983, p. 29). Such a process was only fitful until the nineteenth century when 'it started to spread vigorously' (Melbin, 1987, p. 14). However, as Schivelsbuch (1988) shows, even so this process of colonisation required a considerable period of adjustment away from natural rhythms. For example, it was only at the end of the nineteenth century that streetlighting became independent of natural rhythms like moonlight: 'early in the twentieth century, public lighting in many cities was still regulated according to moonlight schedules. On clear moonlit nights lanterns were turned off earlier than usual, most shortly after midnight' (Schivelsbuch, 1986, p. 91).

A second change was that very gradually a separate set of human practices evolved which we now call 'night life'. In 1738 Hogarth had produced an engraving called 'Night' as a part of a series, 'The Four Times of Day', which showed only lurching drunks and wandering vagrants on a London street. By the eighteenth century lighted pleasure gardens had started to produce a specific night social life, and through the eighteenth century this new kind of sociality was extended by the addition of theatres, shops (which used manufactured light to enhance their displays), cafés and so on. It is no surprise, then, that by the late-Victorian period London at night was seen by many foreign visitors as almost another country:

> Great city of the midnight sun
> Whose day begins when day is done.
> (Le Galliene, cited in Briggs, 1992, p. 24)

As the veil of night was lifted, so many effects were produced. Lighting removed some of the dangers that had once lurked in the dark. Again, the advent of lighting in the household produced new timings and forms of social interaction (Schivelsbuch, 1986; Garfield, 1990). Industrial production was also revolutionised as manufacturers began to work out shift systems which used light and night to enable continuous production.

A third change was that manufactured light was of central importance in the cultural construction of the 'dream spaces' of the nineteenth century (Benjamin, 1973; Buck Morss, 1989); the department stores (with their illuminated display windows), the hotels, the theatres, the cafés, the world fairs, and so on. These urban spaces, which transformed consumption into a mode of being, depended on visual consumption which, in turn, depended upon artificial lighting. Such spaces rapidly became *the* urban experience and their 'spectacular lighting . . . quickly became a central cultural practice' (Nye, 1990, p. 383) providing the opportunity for a new nocturnal round.

A fourth change was that manufactured light produced major new opportunities for surveillance. Foucault has stressed the importance of light to practices of surveillance. There is no hiding in the light. For example, in a prison 'the strong light and the stare of a guard are better captors than the dark, which formerly was protective. The visibility is a trap'. With a light source, a focusing cell and a directed looking all becomes visible (Batchen, 1991). (Similarly, Bachelard (1961, p. 16) notes the importance of light as a means of identification. 'Everything that casts a light sees'.) Manufactured light, then, extended the state and industry's powers of surveillance. The unease that was current in the nineteenth century about street lights that made the inside of houses visible, and the corresponding use of heavy curtains to block out this light (and the gaze of others), help to make this point (Schivelsbuch, 1986).

A final change produced by manufactured light was a remetaphorisation of texts and bodies as a result of the new perceptions. There were, to begin with, new perceptions of landscape: the night under artificial light was a new world of different colours and sensations, modulated by artificial lighting, which painters like Joseph Wright of Derby were amongst the first to explore (see Williams, 1990). Late Victorian London was often seen anew through the power of gaslight:

> London, London, our delight
> Great flower that opens but at night.
> (cited in Briggs, 1989, p. 25)

As importantly, manufactured light began to be used to produce new image technologies and associated institutional forms. In particular, inspired by the camera obscura, a set of technologies began to be produced which prefigured the photograph, such as the phenakisticope and the stereoscope (Batchen, 1991). These techniques were both cause and effect of a reorganisation of the apparatus of vision that had been produced by a new kind of normalised visual knowledge which involved an autonomisation or freeing-up of sight, divorcing it from the necessity to refer to specific spatial locations.

> What begins in the 1820s and 1830s is a repositioning of the observer, outside of the fixed relations of interior/exterior presupposed by the camera obscura and into an undemarcated terrain on which the distinction between internal sensations and external signs is irrevocably blurred . . . there is a freeing-up of vision, a falling away of the rigid structures that had shaped it and constituted its objects.
> (Crary, 1990, p. 24)

This new visual knowledge came to fruition in the photograph (literally light writing) and the serially produced image which can be industrialised; potentially identical objects are produced in indefinite series. On the one hand, the photograph signalled a liberation of the observer. It produced a new regime which permitted new types of image, new forms of fantasy and desire, new forms of 'experience which (did not need) to be equated with presence' (Game, 1991, p. 147). The mediation of the photograph was itself constitutive (Barthes, 1981; Game, 1991). (Other image technologies also heightened this sense of an absent presence, often through contrast between light and dark. Schivelsbuch (1988) shows how inventions of the early nineteenth century like the panorama (circa 1820) and the diorama (1822), and later the magic lantern and film, produced images that traded on the play between light and dark, day and night. 'The power of artificial light to create its own reality only reveals itself in darkness. In the dark, light is life' (Schivelsbuch, 1988, p. 221).) On the other hand, a plurality of institutional means were found 'to recode the activity of the eye, to heighten its productivity and to prevent its distraction' (Crary, 1990, p. 24). No less than Foucault's panopticon, nineteenth century image technologies were surrounded by institutions of surveillance which

> invoked arrangements of bodies in space, regulations of activity, and the deployment of individual bodies, which codified and normalised the observer within rigidly defined systems of visual consumption. There were disciplinary techniques for the management of attention, for imposing homogeneity, anti-nomadic procedures that fixed and isolated the observer.
>
> (Crary, 1990, p. 18)

2.1.3 Power

A further important machinic complex is power. Electrical power is, of course, hardly a new invention. The invention of the Leyden jar sparked off the first electrical experiments in Germany and Switzerland in the 1740s. By the 1780s electricty had achieved some measure of recognition in medical circles, a popularity which it retained in the nineteenth century through devices like galvanic belts, electric baths, electric shock treatment and the like, for the treatment of physical and mental disorders (Armstrong, 1991; Nye, 1990). The telegraph reinforced this recognition by bringing about changes in 'the nature of language, of ordinary knowledge, of the very structures of awareness' (Crary, 1989, p. 202). It gave a particularly strong boost to a rapidly expanding 'rhetoric of the electrical sublime' (Carey, 1989). This rhetoric was particularly marked in the United States where it tended to be couched in the language of either religious aspiration, or secular millenarianism. Henry Adams, for example, saw the telegraph as a demonic device displacing the Virgin with the Dynamo. But for Samuel Morse, the telegraph was an electrical nervous system able 'to diffuse with the speed of thought a knowledge of all that is occuring through the land' (cited in Carey, 1989, p. 207). In Britain, the rhetoric of the telegraph tended to be less extreme but, even there, it fired a powerful flow of thought, most especially when connected with matters of Empire for which the telegraph was a powerful integrative metaphor.

But it was the promise of infinite light and power that made electricity into a household term. In particular, it was the advent in the 1880s of power stations which distributed electricity along a network that allowed electricity to come into its own, rather as gas had some decades earlier. Indeed, at first, these power stations were often depicted as electrical gas works. But as it became clear that high voltage could be transported over long distances, electricity power stations were located at long distances away from where the electricity was actually used, thus giving the impression that electricity was a sourceless source, an absent presence (Schivelsbuch, 1986). Further, unlike gas, electrical light was immediately accepted in drawing rooms as clean and odour free. Electricity therefore reached more places more quickly and it quickly became a vast *collective* network of power.

Already in 1881 *Punch* was able to introduce into its pages a cartoon showing King Coal and King Steam watching the infant electricity grow and asking 'what will he grow in to?' Certainly, in the 1880s, it looked as though London might become an 'electropolis'. By 1891 London had 473,000 incandescent lamps, compared with Berlin's 75,000 and Paris's 67,600 (Hughes, 1983). But in the 1890s rates of electricity adoption in Britain slowed markedly – the result of a combination of recession, government legislation and opposition from gas interests – and never really recovered until the 1920s. By 1913 Berlin and Paris had more generating capacity than London, partly the result of having a few large generating stations compared with London's many local ones. Similarly, electricity seems to have failed to capture the British imagination in the same way that it did that of the United States or some countries of Europe where popular absorption in the potentiality of electricity as a force for personal and social translation was much greater (Nye, 1990), and the mystery and magic of electricity was more appreciated (Briggs, 1992). In part, this seems to have been the result of the much greater purchase that romantic notions of an arcadian countryside and nature mysticism were able to get on the British imagination with an associated rejection of 'mechanisation' in all its forms (Naylor, 1992).

Like speed and light, electrical power produced a number of important changes. Three of these stand out. The first change was the boost electrical power gave to new conceptions of time and space. Through the telegraph and the instant power provided by the light switch came ideas of absent presence, of a geography of communication that no longer depended upon actual transport (Carey, 1989). A second change was the remetaphorisation of the body into the body electric. From the time of *Frankenstein*, 'electricity, energy and life (had been) synonymous' (Schivelsbuch, 1988, p. 71). Now forces, engines, dynamos, discharges, currents and flow become a new linguistic currency of desire. More than even speed or light, electrical power

> raised the possibility of an alteration in what Felix Guattari calls the 'social chemistry of desire', in the means through which physical desire, and motivation generally, could be represented, articulated and transmitted. If the social and the individual body are intimately related, then the massive programmes for the wiring of individual houses that went on throughout the late nineteenth century have a parallel in the wiring of individual bodies and in the way in which states of desire, animation, incandescence are figured in literary and other texts.
>
> (Armstrong, 1991, p. 307)

Thus a whole vocabulary grew up (live wires, human dynamos, electric performances, and so on) (Nye, 1990) that depended upon electricity, conceived as 'shadowy, mysterious, impalpable. It lives in the skies and seems to connect the spiritual and the material' (Czitrom, cited in Carey, 1989, p. 206).

The third and final change was electricity's critical role in the change to a new form of more integrated capitalism. Society was tied together and made subservient to the new networks of power:

> the transformation of free competition into corporate monopoly capitalism confirmed in economic terms what electrification had confirmed technically; the end of individual enterprise and an autonomous energy supply. It is well known that the electrical industry was a significant factor in bringing about these changes. An analogy between electrical power and finance capital springs to mind. The concentration and centralisation of energy in high capacity power stations corresponded to the concentration of economic power in the big banks.
>
> (Schivelsbuch, 1988, p. 74)

2.2 The twentieth century

2.2.1 Speed

In the twentieth century, the history of speed has been a continuation by other means of methods of transport and communication that started in the nineteenth century. The automobile has, perhaps, been the most important device. It is often viewed as the avatar of mobility. For Baudrillard (1988, pp. 52–3), for example, Los Angeles freeways speak to a society built around the automobile and the 'only truly profound pleasure, that of keeping on the move',

> Gigantic, spontaneous spectacle of automobile traffic. A total collective act, staged by the entire population, twenty four hours a day. By virtue of the sheer size of the layout and the kind of complicity that binds this network of thoroughfares together, traffic rises here to the level of a dramatic attraction, acquires the status of symbolic organisation. The machines themselves . . . have created a milieu in their own image.
>
> (Baudrillard, 1988, pp. 52–3)

But in Britain, in contrast to the United States, the automobile's status as the prime mover was hampered by the British class system. Private cars and vans on the road increased from 132,000 in 1914 to nearly 2 million in 1939 (Thrift, 1990; Robson, 1990) but 'possession of a car still remained exclusive since those numbers represented only a very small proportion of the population' (Robson, 1990, p. 67). It was not until the late 1950s and early 1960s that car ownership started to become general as real car prices fell dramatically. Car ownership increased from under 4 million in 1956 to 6.5 million in 1962 and 10 million in 1967, but even so, there were dramatic spatial variations. By 1966 there were 3.8 people per car in Surrey but still only 11.2 in Glasgow Coatbridge (Plowden, 1971). The general increase in access to speed at this time was also underlined by the building of the motorway network which started with the opening of the 8-mile Preston bypass in 1958 (which was to become a part of the M6) and the first 72 miles of the M1 in

1959.[12] Britain can now lay claim to being a mobile society in the North American sense but even in the 1990s car ownership levels have reached no more than 403 cars per 1,000 people, compared with 589 per 1,000 in the United States. (However, it is also worth remembering that in Britain the motor bus, which first came into operation in 1903, has had a compensating and largely unsung history as a mover of people.)

Again, unlike the United States, the aircraft did not come into general use as a mass passenger carrier until comparatively late on. Although passenger services were instigated in the 1920s it was not until the 1970s and the advent of package tourism to Europe, that the aircraft became a general means of travel. However, it is important to note that the aircraft had had a much more general impact on the popular imagination long before it came into general use both as a symbol of modernity and as a source of new perspectives.

New networks of instantaneous communication were also coming into being gradually replacing the telegraph. Bell's telephone was first exhibited in Britain in 1877. Although exchanges had opened in most major cities by 1882 (beginning in London in 1879) and an extensive network had developed by 1892, this did not signify any mass take-up of the telephone. Even in the 1880s, London had fewer subscribers than most major North American cities, or even Paris. By 1913 only 663,000 phones were in use in the whole of Britain (Thrift, 1990). Indeed, usage of telephones was restricted to the middle and upper classes until well after the Second World War; it was not until the 1970s that telephone ownership began to become general and not until the 1980s that more than three-quarters of British households had access to a telephone. Further, long-distance telephony was relatively unusual for many decades, partly because of the local origins of the British telephone system. This slow diffusion of telephones was in marked contrast to the United States and was chiefly the result of different, more class-based perceptions of the telephone, the monopoly powers of the Post Office, and the efficiency of the mail and telegraph systems (De Sola Pool, 1977).

2.2.2 *Light*

In the United States electrical lighting was at first used chiefly in downtown shops and streets; 'the streets of the downtown entertainment districts . . . were by 1900 literally bathed in electric light' (Nasaw, 1992, p. 275). But residential areas were still usually lit by gas. It was not until after 1910 that electric lighting started to be used on a wide scale in urban areas and by 1930 'the majority of urban dwellers had electric lights' (Nye, 1990, p. 303) (although it was not until later still that rural electrification really began to bite). What is clear is that from an early date electric lighting had become a standard element

[12] But Virilio (1991a) suggests that such a claim can be disputed. Whereas the freeway system can be regarded as a place of integration, doubling back on itself, the motorway system still chiefly consists of unique directional axes, intended to expel traffic. It is also important to remember, in this context, the role of the railway network in Britain.

of the North American urban experience, especially through street lighting and the use of electric lighting to create urban spectacles such as the Great White Way in New York (Nye, 1990). Thus, the novelty of electric lighting quietly faded.

In Britain, by contrast, the progress of electric lighting was clearly slower, hampered by the lack of electrical power. For example, in Manchester in 1914 only one in fifteen of the inhabitants of the administrative area supplied by the Manchester Electricity Department lived in houses which were either heated or lit by electricity. By the mid-1920s this figure had risen to one in seven and by the mid-1930s to one in two. These figures were probably good, compared with the nation as a whole. For example, as late as 1938, in many Welsh rural areas, wax candles and paraffin were still the main forms of lighting and in some remote Welsh rural areas rushlights and tallow-tip candles could still be encountered (Luckin, 1990).

Other new inventions based on electricity and light had a more dramatic impact and, most particularly, the cinema. In 1894 Edison unveiled the kinetoscope in Britain, and in 1896 a cinematograph was set up in London. By 1914 the cinema was already a major industry with somewhere between 4,000 and 4,500 picture houses and audiences counting in the millions. Cinema attendance reached a peak in 1939, a year in which 5,000 picture houses sold 20 million tickets a week (meaning that about 40 per cent of the population went to the cinema at least once a week, and about a quarter twice a week) (Robson, 1990). The invention of television took some of the characteristics of cinema into the home. Transmissions began in Britain in 1936 but it was not until television was revived after the Second World War, in 1946, that it began to become popular as a cultural medium. In fact, at first the number of licenses grew slowly, to 45,564 (in a population of approximately 50 million) by 1948. The catalyst that sparked mass ownership was clearly the Coronation of Elizabeth II in 1953. Over 20 million of the population watched this event on television, nearly 8 million in their own homes and well over 10 million at the homes of friends (Briggs, 1985). By 1956, the year when commercial television first started broadcasting (Sendall, 1982) and broadcasting hours were significantly extended, 'the ownership or rental of a television set was passing through and out of the stage of being a matter of status' (Corner, 1991, p. 6). Certainly, by the late 1960s, when over 16 million licenses were issued, television had become the principal instrument of public information and cultural identity.

2.2.3 Power

Finally, there is the history of power. In many parts of the world, electricity took on an almost mythical status (Carey, 1989). As a symbol of progress and modernity 'electricity seemed to ensure a brilliant future for civilisation. Authors in the popular press referred to it as white magic that promised an electrical millenium' (Nye, 1990, p. 66). Cities like Berlin, New York and

Chicago with their enormous power-generating capacity were thought of as 'electropolises'. Further, between 1880 and 1920 electrical power became increasingly important worldwide as the source of power for local traffic systems, the elevator, the radio and the cinema and, increasingly, elaborate lighting displays.

However, in Britain the adoption of electricity as a routine source of power was actually quite slow:

> domestic use of electricity made no rapid advance in England, partly because only about two per cent of homes were connected to the mains in 1910 and partly because the cost remained high; a number of progressive power undertakings . . . were just beginning to cut charges from 8d to 1/2d a unit in 1914 when war intervened.
>
> (Burnett, 1986, p. 214)

Even in 1931, one article reported that 'of 10½ million houses in this country, less than 30 per cent are wired, and not 1 in 1,000 is all-electric. There are nearly 8 million houses to go for, most of which are either within reach of distributing mains or in areas in which main-laying would prove profitable' (Luckin, 1990, p. 52).

As a result electrical domestic appliances also spread slowly, especially compared to the United States (Hayden, 1981; Tichi, 1987). Thus Britain's 'grand domestic revolution' (Hayden, 1981) was, in effect, postponed until after the Second World War. Part of the reason for this slow domestic spread was that coal and gas remained cheap and easily accessible resources. Part of it was that domestic servants were still used to a much greater extent than in other countries like the United States, thus negating the need for many middle-class women to turn to 'time-saving' devices until the 1920s and 1930s. And part of it was the lack of constant central power supplies. This latter problem began to be solved in 1926 by the enacting of a bill to establish a National Grid. Construction of the Grid ended in 1933 and by 1936 the Grid was in full operation in all regions except the north-east of England (Hughes, 1983). The Grid both brought down the price of electricity and gave an enormous boost to British industry. But it was not until the period between 1945 and 1950, and nationalisation of the electricity industry, that 'electricity established itself in a growing number of urban areas and penetrated more deeply into hitherto poorly served country districts' (Luckin, 1990, p. 172).

None of this is to suggest that there were no advocates of the electrical sublime in Britain. There clearly were. Luckin (1990) has documented the rise of an 'electrical triumphalism' and the cult of electrical progress that sometimes seems every bit as pugnacious as its North American counterpart, although often with its own nuances (for example, a tendency to visions of a technological arcadia which would 'stimulate a second and cleaner industrial revolution, the decentralisation of mass production and a regeneration of a deeply depressed agrarian society' (Luckin, 1990, p. 11). Further, this triumphalism was widely promoted, especially, in a domesticated form, to housewives through advertisements, journals and the Electrical Association for Women. But the opposition to the encroachment of electricity, especially in the form of pylons and power lines, was much greater, often predicated on romantic themes of neo-arcadianism and nature mysticism that had originated in the previous century, and before. The power of these discourses was sufficient to affect the enthusiasm for electricity in Britain.

These changes in the machinic complexes of speed, light and power were, by themselves, remarkable. However, I think it would be possible to argue at rather greater length than I will here that their effects were essentially a continuation of tendencies set in train in the nineteenth century.

So far as the machinic complex of speed is concerned, there seems little reason to believe that consciousness of time and space altered radically from the later nineteenth century (Kern, 1983). Of course, the population did pay more attention to ever-finer durations of time, a necessary skill to deal with ever-finer timetables, but this was a quickening, not a qualitative change. Again it might be argued that panoramic perception was strengthened (especially by the invention of the cinema) rather than altered. The effects of the general speed-up on texts also continued apace (often taking on extreme forms as in Futurism), as did the modernisation of the subject. Metaphors of progress and circulation continued to abound.

So far as the machinic complex of light is concerned a similar continuity between the late nineteenth and the twentieth century can be found. The spread of artificial lighting and the colonisation of the night continued apace.[13] In J. B. S. Haldane's words, 'the alternation of day and night is a check on the freedom of human activity which must go the way of other spatial and temporal checks'.[14] Perceptions of the new landscapes of night space – Virilio's (1991b) 'false day' – created by artificial light also began to stabilise in at least three ways. First,

> electrification made possible a new kind of visual text, one that expressed an argument or view of the world without writing, solely through suppressing some features of a site and expressing others. The new rhetoric of night space edited the city down to a few idealised essentials. It underlined significant landmarks and literally highlighted important locations. By night, the spectators could grasp the city as a simplified pattern. The major streets stood out in white bands of light, the tall buildings shone against the sky, and other important structures such as bridges hung luminously in the air, outlined by a string of bulbs. The city

[13] None of this is to suggest that night still does not hold its terrors, especially for women, as the numerous marches to regain the night attest.

[14] In Britain, a large night-time labour force came into being as did numerous incessant economic and social activities. By the 1970s 14 per cent of the UK workforce was doing shiftwork. In 1978 a 24-hour radio broadcasting service was set up and in the late 1980s all-night television broadcasting began. Again, in the late 1980s 24-hour shops on the American model became more and more common. (These figures cannot compare with those of the United States where over 29 million people were active after midnight in 1980, where 24-hour radio broadcasting first started in the 1960s, where 24-hour television broadcasting started in the 1970s and where there are now many 24-hour shops).

> centre blazed its importance . . . and the corporations erected electrical signs to proclaim their products and their importance.
>
> (Nye, 1990, p. 60)

Second, cultural workers like visual artists began to represent this new world in their work, often as a kind of 'electrical sublime'. As the work of artists like Hopper and O'Keefe attests, 'most of the important painters of the early twentieth century attempted to paint the city at night' (Nye, 1990, p. 76). Third, the city at night began to be perceived as just as real, or even more real, as the city in the daytime: lighting served to recover streets and buildings not only from dimness but also from banality (Nye, 1990, p. 61). Lighting became essential to glamorise the city; even poorer areas could take on a different kind of nuance (Nye, 1990). Thus night and light became intertwined. Not surprisingly Benjamin's urban phantasmagoria of dream spaces was only strengthened by these kinds of developments. The intensity of surveillance resulting from the widespread use of artificial light has also grown.

However, perhaps, the most important developments in the effects of the machinic complex of light emanated from the development of cinema, and then television and video, the basis of the industries of the image and spectacle that now dominate our culture, whose images have, in effect, become the 'raw material of vision' (Virilio, 1988, p. 66).[15] The most notable of these developments was clearly the simple speed-up of the image:[11] 'the very name of the new medium identified its effects – moving pictures' (Kern, 1983, p. 117). Early viewers were fascinated by any moving image, 'by the speed effects of light' (Virilio, 1988, p. 32) and subsequent advances in editing added to the breathlessness of many presentations. Another effect was the new ways in which the passage of time could be conceived as a result of movement and editing;

> Chronological development gradually did away with the longeurs of the old photographic pose; the architecture of the set, with its spatial mass and partitions, supplanted free montage and created a new emphasis. Rather like my grand-daughter, who, when she moved from one room of her flat to another, used to think that a different sun was shining into each one, so the cinema marked the advent of an independent and still unknown cycle of light. And if it was so hard for the photographs to move, this was above all because the operation of moving cinematic time – of preserving its original speed in an old, static and rigidly ordered environment – was as astonishing for these early pioneers as it was difficult to invent.
>
> (Virilio, 1988, p. 13)

Finally, cinema produced an 'aesthetics of astonishment' by making visible something which could not exist otherwise (Vasseleu, 1991), through a process of recognising the familiar; 'opening up new possibilities for exploring reality and providing means for changing culture and society along with those

[15] There is no space here to note, in particular, the effect of advertising as a part of the developing technology of light. But see Taussig (1993).

possibilities. . . . All this is summed up in Benjamin's notion of the camera as the machine opening up the optical unconscious' (Taussig, 1993, pp. 23–4). However, it is a moot point whether the advances in photography signalled by the cinema and then television actually constituted a major shift. One might even question whether the video, with its ability to 'time shift' (Cubitt, 1991) and its natural supplementarity, signifies a new set of cultural relations. Each of these different inventions is still effectively a mimetic medium that corresponds to the optical wavelength of the spectrum and to the point of view of a positioned observer located in real space, just as the photograph does (Crary, 1990).

So far as the machinic complex of power is concerned, a similar continuation of its effects can be found from the nineteenth century. Instantaneity was extended as electrical and communication networks continued to burgeon. The re-metaphorisation of the body as an electrical flow was continued (see Armstrong, 1991). Electricity continued to galvanise the practices of capitalism. By the middle of the twentieth century, the modern city was 'inconceivable without electricity' (Ward and Zunz, 1992, p. 11).

However, it has also been shown that the three machinic complexes of speed, light and power had quite different histories in different spaces. Using the example of Britain, it has been possible to show a quite different trajectory in a number of cases. Technocracy and fordism never took on the extreme forms typical of North America and, in different guise, other parts of Europe (Ross, 1991). In North America, functionalist streamlining, prominent machinescapes, and the work of novelists like Dreiser, Norris, London or Williams (Tichi, 1987) and artists like Sheeler (Lucic, 1991), gave shape to the values of machine energy. In continental Europe, Bauhaus, Futurism (and especially Marinetti's celebration of the motor car), celebrations of 'Americanism' (Wollen, 1993), and Le Corbusier's vast utopian structures did the same. In Britian, in partial contrast, these values never came to a full flowering (but see Forty, 1986).

2.3 The late twentieth century: mobility

I want to argue that, beginning in the 1960s in most western countries, it becomes possible to begin to talk about a new synthesis of speed, light and power (which I have called mobility) resulting from the growth of what can be called 'active' machinery. No longer is it possible to see the human subject and the machine as aligned but separated entities, each with their own specific functions. No longer it is possible to conceive of the human subject as simply 'alive' and the machine as just so much 'dead labour', the two linked only by their capacity for movement.

This realisation is prompted by radical changes in the machinic complexes of speed, light and power. First, the machinic complex of speed begins to transmute. In particular, there has been a massive increase in the volume of travel in and between western cultures. This is coincidental with the general democratisation of the automobile and the aircraft and is perhaps best illustrated by the case of mass tourism. By 1989 there were 400 million international arrivals a year, compared with 60 million in 1960. There were between three and four times this number of domestic tourists worldwide (Urry, 1990). Further,

> much of the population will travel somewhere to gaze upon it and stay there for reasons basically unconnected with work. Travel is now thought to occupy 40 per cent of available 'free time'. If people do not travel they lose status: travel is the maker of status. It is a crucial element of modern life to feel that travel and holidays are necessary. 'I need a holiday' is the surest reflection of a modern discourse based on the idea that people's physical and mental health will be restored if they can get away from time to time.
>
> (Urry, 1990, p. 5)

Communication technology has also been the subject of a surge with a consequent massive increase in the volume of indirect communication. Satellite communication (Jayaweera, 1983) and the fibre optic cable, in conjunction with the computer, have revolutionised telephony which has metamorphosed into 'telecommunications', involving not just voice messages but also high-speed data transfer, fax, and the like. The advent of the mobile telephone is likely to lead in time to direct person-to-person communication with no time lags.

It might just conceivably be argued that these changes are simply a continuation by other means of nineteenth century themes. The same thought cannot be applied to the complexes of light and power. Thus the machinic complex of light has seen a virtual transformation, symbolised by the computer, the fibre optic cable, the laser and the increasing use of the non-visible spectrum expressed in technologies like computer-aided design and graphics, synthetic holography, flight simulators, computer animation, robotic image recognition, ray tracing, texture mapping, motion control, virtual reality helmets and gloves, magnetic resonance imaging, multispectral sensors and medical imaging (Crary, 1990; Vasseleu, 1991). As Wollen (1993, p. 64) puts it,

> the invention of the computer . . . has transformed the entire field of image production. We are now in the first phase of video-computer productions. In the 1920s Dziga Vertov described the camera as a mechanical eye; now it is the mechanical eye of an electronic brain. Indeed, the camera itself has been transformed. It is now simply one option within a whole range of sensors and information-recording devices, some visual, some non-visual. Images can be produced by means of x-rays, night vision, thermal, magnetic, electronic spin and a host of other kinds of sensors. The camera (film or video) is simply the one that still most closely appropriates 'natural vision'. Sensors are no longer hand-held or mounted on tripods with humans attached to them, looking down view finders. Their motion and action can be remote-controlled. They can circle the earth in satellites and transmit from within the human body.
>
> (Wollen, 1993, p. 64)

These new image technologies have three consequences. First, they relocate vision to a plane severed from a human observer;

> Most of the historically important functions of the eye are being supplemented by practices in which visual images no longer have any reference to the position of a observer in a 'real' optically perceived world. If these images can be said to refer to anything, it is to millions of bits of electronic mathematical data. Increasingly visuality will be situated on a cybernetic and electromagnetic terrain where abstract visual and linguistic elements coincide and are consumed, circulated and exchanged globally.
>
> (Crary, 1990, p. 2)

Second, they have produced, in conjunction with radar, radio and microwave technologies, new possibilities for surveillance. Delanda (1991) writes of the growth of new systems of 'panspectric' surveillance that are replacing panoptic systems, in which the visual is redefined to include more of the spectrum, both up and down, of a new non-optical (or rather extended optical) intelligence-acquisition machine based on 'dark light'.

Third, the relation of the subject to these new technologies is no longer the relation of the active to inactive. Increasingly the subject scans a real-time simulated field of action of which the body (or the observing body) is a component. Thereby, 'thereness' is redefined. This fusion of human and machine has gone farthest in the military sphere (Delanda, 1990):

> the fusion is complete, the confusion perfect; nothing now distinguishes the functions of the weapon and the eye; the projectile's image and the image's projectile form a single composite. In its tasks of detection and acquisition, pursuit and destruction, the projectile is an image or signature on the screen, and the television picture is an ultrasonic projectile propagated at the speed of light.
>
> (Virilio, 1988, p. 83)

In turn,

> the disintegration of the warrior's personality is at a very advanced stage – looking up s/he sees the digital display (opto-electronic or holographic) of the windscreen collimator; looking down, the radar screen, the onboard computer, the radio and the video screen, which enables him to follow the terrain with its four or five simultaneous targets and to monitor his self-navigating Sidewinder missiles, fitted with a camera or infra-red navigating system.
>
> (Virilio, 1988, p. 84)

Finally, there is the machinic complex of power. Here again enormous change has been generated, symbolised by the microchip and the computer. The history of the microchip and computer and cybernetics is now so well known that it hardly bears repeating. More important has been its recent history, and especially the increasing importance of software compared with hardware, and the consequent rise of 'artificial' intelligence. However, most importantly, as artificial intelligence has grown as an area of inquiry so it has started to cast off the anthropocentrism of its earlier history, and has moved towards a view that artificial (or infra-human) intelligence is not a simulation of human intelligence but has to be understood as an 'other' kind of intelligence (Dreyfus, 1992), as 'liveware'.

> it emphasises the commonality of human and infra-human intelligence on the one hand, and on the other hand it inspires a respect for infra-human aspects of intelligence by illuminating their immense sophistication and complexity as against the relative simplicity and mechanical reproductivity of the uniquely human aspects.
>
> (Preston, 1992 p. 14)

That these three different machinic complexes are now in the process of merging is nowhere better illustrated than by the case of the melding of the computer with old and new image recording technologies (Wollen, 1993). The new integrated systems already include the following characteristics: access to a database of stored images in an electronic memory, with all the consequent possibility of combination with images from different sources; generation of images and text by computer; simulations of the 'real world'; hybrid imaging; and connection with new modes of transmission and reception. In turn the possibilities arising out of this melding open up new problems, especially the appropriate forms of logic and aesthetic.

> The first requirement is the development of a heterogenous theory of meaning, open rather than closed, involving different types of sign, and bringing semantics together with hermeneutics, reference with metaphor. The second is a specific (formal) theory of intertextual meaning, the way in which re-contextualisation changes meaning, the double hybrid coding involved in quoting, plagarising, grafting and so on, the back and forth of meaning between texts.
>
> (Wollen, 1993, p. 67)

The changes in these three, once different machinic complexes begin to offer us glimpses of a 'cyborg' culture of mobility in which machines constantly pose the problem 'what counts as a unit' (Haraway, 1991a, p. 212);

> Modern machines are quintessentially microelectronic devices. They are everywhere and they are invisible. Modern machinery is an irreverent upstart god, mocking the father's ubiquity and spirituality. Writing, power and technology are old partners in western stories of the origin of civilisation, but minaturisation has changed our experience of *mechanism* . . . our best machines are made of sunshine; they are all light and clean because they are nothing but signals, electromagnetic waves, a section of a spectrum, and these machines are eminently portable, mobile. . . . People are nowhere near so fluid, being both material and opaque. Cyborgs are ether, quintessence.
>
> (Haraway, 1991a, p. 153)

Cyborg culture is no longer concerned with mimetic, analog media (that constitute a point of view) but is establishing its own active actant reality (Poster, 1990). In the process the human subject is being modernised so that it is able to cope with this new order of reality in which machines are as active as human subjects, in which there is a very high degree of interactivity between machines and humans, and even in which, in certain cases, machines become parts of the human subject's body (as for example, in the case of 'smart' materials like artificial pancreatic cells and menotic implants that integrate with the host tissues (Coghlan, 1992)). There are, in other words, both new kinds of bodies and a technologising of the human body, events which are marked by a refiguring of human potential and an increasingly abstract awareness of the body as network and system (Armstrong, 1991). Thus it now becomes possible to consider a new relationship between humans and machines and machinery and humans which has been described by some commentators (e.g. Mandel, 1978; Jameson, 1991) as the Third Machine Age;

> if motorised machines constituted the second age of the technical machine, cybernetic and informational machines form a third age that reconstructs a generalised regime of subjection; recurrent and reversible 'humans–machines' systems replace the old nonrecurrent and nonreversible relations of subjection between the two elements: the relation between human and machine is based on internal, mutual communication and no longer on usage or action.
>
> (Deleuze and Guattari, 1980, p. 3)

Now, this apocalyptic style of writing is clearly open to abuse. It can lead to Boys Own hyperbole of the worst kind;

> It is ceasing to be a matter of how we think about technics if only because technics is increasingly thinking about itself. It might still be a few decades before artificial intelligences surpass the horizons of biological ones, but it is utterly superstitious to imagine that the human dominion of terrestrial culture is still marked out in centuries, let alone in some metaphysical perpetuity. The high road to thinking no longer passes through a deepening of human cognition, but rather through a becoming inhuman of cognition, a migration of cognition out into the emerging planetary technosentience reservoir.
>
> (Land, 1992, p. 218)

But it is not necessary to go all the way towards all-seeing 'fantasies that the world might run without human interaction' (Strathern, 1992a, p. 169). It is possible to think about the Third Machine Age in another, more prosaic way (Collins, 1990). Our relationship to machines is changing for three reasons. First, machines have got better at mimicking us through more complex and rarified behaviour. Second, we have become more 'charitable' (Gellner, 1974) to the strangeness of machines, both in sense that we are more likely to welcome 'mechanical strangers' (Collins, 1990) and in the sense that we are more likely to behave like machines ourselves. For example, we might adjust our writing style to the new machines, as Carey (1989) documents for the case of the telegraph and press reporting in an earlier era. Third, our image of ourselves becomes more like our image of machines. For example,

> Man's nature is indeed so malleable that it may be on the point of changing again. If the computer paradigm becomes so strong that people begin to think of themselves as digital devices on the model of work in artificial intelligence, then, since for the reasons we have been rehearsing,

> machines cannot be like human beings, human beings may become progressively like machines. During the past two thousand years the importance of objectivity, the belief that actions are governed by fixed values; the notion that skills can be formalised; and in general that one can have a theory of practical activity, have gradually exerted their influence in psychology and in social science. People have begun to think of themselves as objects able to fit into the inflexible calculations of disembodied machines.
>
> (Dreyfus, 1979, cited in Collins, 1990, p. 189)

What seems clear is that our current ways of 'representing' a more mobile Third Machine Age, even in a form that is rather less dramatic and contains greater elements of continuity with the past than Jameson or Deleuze and Guattari might acknowledge, are not equal to the task. In part, this is a matter of the newness of much of the relevant machinery which has not yet had time to become culturally 'bedded in'. In part, it is also because of the nature of the new machinery which does not lend itself to conventional representations. Thus to a degree at least, it is possible to agree with Jameson (1991, p. 45) that one problem is that this new active, actant machinery, unlike older forms of machinery like the railway engine or the aircraft or even the electrical turbine, does not 'represent motion but can only be represented *in motion*'. Much the same problem occurs in the case of the depiction of new actant subjects which are *hybrids*, humans *and* machines in actant couplings.

This is why the notion of structure of feeling is so crucial. It is able to capture the importance *and* 'thisness' *and* liminality of much new experience and, at the same time, the difficulty of representing it in theory and in practice. The notion also suggests that in particular periods a general change of style is needed. It is this change of style which is the subject of the next two sections of this chapter.

3. THE UNSETTLING. TRAVELLING BETWEEN HISTORY AND GEOGRAPHY

This brief history of the development of the structure of feeling that I have called mobility is, I would claim, a necessary prologue to a consideration of current theoretical debates in the social sciences and humanities around poststructuralism, postmodernism and postcolonisalism. In particular, I would claim that the new machinic world of 'technosentience' (Land, 1992) saturates these debates, providing much of the vocabulary in which they are couched and, more importantly, a distinctive (but not distinct) ontology. Put at its crudest, this ontology is a description of the conditions of questioning modern life. Thus its chief concern is a series of root related conditions which are endemic to modern life. The first of these is *flow*, 'flows of energy that irradiate, condense, intersect, build, ripple' (Lingis, 1992, p. 3). The second condition is *networks*. As Mulgan (1991, p. i) has pointed out:

> The late twentieth century is covered by a lattice of networks. Public and private, civil and military, open and closed, the networks carry an un-

> manageable volume of messages, conversations, images and commands. By the early 1990s the world's population of 600 million telephones and 600 million television sets will have been joined by over 100 million computer workstations, tens of millions of home computers, fax machines, cellular phones and pagers. Costs of transmission and processing will continue their precipitate fall.
>
> (Mulgan, 1991, p. 6)

Societies act to record, channel and regulate the flow of energies. The third condition is *power*. The modern world is under the sway of powerful forces like capitalism and the state whose effectiveness depends on their ability to mobilise resources in war-like power struggles which reach down even to the capillary level. This power is both potential and determined (Massumi, 1992). The fourth condition is *boundaries*. The modern world is increasingly seen as decentralised and fragmented, whether the subject is states, capital, bodies, machines – or subjects. Space therefore takes on critical importance as both a battlefield and a zone of mixing, blending, blurring, hybridisations. The fifth condition is *absence*. How is it possible to understand a world in which the event can no longer exist independently but is filled with the echoes of other events pulsing through the network, in which every attempt to make a point is foiled by its own traces? In other words, presence, perspective and the whole visual metaphor on which so much knowledge is based become suspect. The final condition is *time*. Time becomes a complex and acclerated multiple phenomenon which respects no master narrative (Adam, 1991).[16] These six conditions constitute a new kind of materiality in that they attest to a concern with a redrawing of the boundaries between what Marx called 'the world of men' and 'the world of things' (Marx, 1844, p. 272), what Veblen called 'the animate and the inanimate' (cited in Seltzer, 1992, p. 3) and what Haraway (1985, p. 153) calls 'the physical and non-physical'. In what follows, I will simply list some of the chief ramifications of this ontology.

(1) There is little sense to be had from making distinctions between space and time – there is only *space-time*. Attempts to privilege either time or space, suggesting that one or the other is the signature of an age, for example, make only limited sense. First of all mobility takes up both space and time, one of the elementary insights of a time-geography now either forgotten or misunderstood (Parkes and Thrift, 1979).[17] Second, even if in some mystical past it had been possible to analytically separate space and time, in the contemporary world, 'the notions of space as enclosure and time as duration are unsettled and redesigned as a field of infinitely experimental configurations of space-

[16] The old Newtonian linear time has, since Freud and Einstein, been replaced by a plurality of times (Balibar, 1972; Foucault, 1973; Parkes and Thrift, 1979; Major-Poetzl, 1979; Wilcox, 1987). Similarly, there has been a pluralisation of space (Lefebvre, 1991).

[17] Time-geography has been chronically misunderstood in recent writings. It serves the purposes of this chapter well for two reasons. First, as an ecological approach, it makes no distinction between things. Hägerstrand was always clear that lifelines should apply equally to machines and animals as to human beings. Second, as an attempt to write mobility, Hägerstrand was always clear that he saw time-geography as an attempt to write a kind of musical score.

time' (Emberley, 1989, p. 756). If there is a view of time and space that this corresponds most closely to, it is probably Bergson's view of time-space as a permanently moving continuity, a qualitative multiplicity (Game, 1991; Deleuze and Guattari, 1988).

(2) In contemporary societies, *mobility* has become the primary activity of existence (Prato and Trivero, 1985). This is a perspective that has a long philosophical history, of course. For example, we can trace a direct line of descent back from modern poststructuralist authors to authors like Nietzsche, Heidegger, Bataille and Blanchot.[18] Running through these authors' work is the hum of mobility. Thus, in Nietzsche there is the conception of the world as a constant becoming of a multiplicity of interconnected forms. It is quite clear that, to an extent, Nietzsche took this idea from the world around him as well as the world of the text. 'Nietzsche speaks of the "haste and hurry now universal . . . the increasing velocity of life", of the "hurried and over-excited worldliness of the modern age"' (Prendergast, 1992, p. 5). For him, the modern individual faces a crisis of assimilation because of the 'tropical tempo' of this modern world.

> Sensibility immensely more irritable; . . . the abundance of disparate impressions greater than ever; cosmopolitanism in foods, literatures, newspapers, forms, tastes, even landscapes. The tempo of this influx *prestissimo*, the impressions erase each other; one instinctively resists taking in anything, taking anything deeply, to digest anything; a weakening of the powers to digest results from this. A kind of adaptation to the flood of impressions takes place: men unlearn spontaneous actions, they merely react to stimuli from the outside.
>
> (Nietzsche, 1967, p. 47).

In Nietzsche's words, 'with the tremendous acceleration of life, mind and eye have become accustomed to seeing and judging partially or inaccurately, and everyone is like the traveller who gets to know a land and its people from a railway carriage'.

This same absorption into this new world can even be found in Heidegger's notion of encountering/letting it show up (Dreyfus, 1991). As Dreyfus (1992) has shown, what in Heidegger is depicted as a primordial way of being actually reflects a profoundly modern experience of encountering equipment, the ground for which is prepared in *Being and Time* and elaborated on in Heidegger's later work on the 'total mobilisation' of all beings.

Even in writers a little closer to the Parmenidian tradition, which reaches its acme in the inexorable workings-out of Hegelian dialectic, there is the same concern with mobility. Simmel, for example, takes up Marx's concerns with the general speed-up in the circulation of commodities and the consequent 'agitation' of modern life and Nietzsche's idea of modern anti-aesthetic distraction as a central feature of the modern world and links them together in his familiar critique of 'impatient' modern urban life;

[18] As well as the other writers like Hobbes and Veblen. See Seltzer (1992), Prendergast (1992).

> It is true that we now have acetylene and electrical light instead of oil lamps; but the enthusiasm for the progress achieved in lighting makes us sometimes forget that the essential thing is not the lighting itself but what becomes more visible. People's ecstasy concerning the triumph of the telegraph and the telephone often makes them overlook the fact that what really matters is the value of what one has to say, and that, compared with this, the speed or slowness of the means of communication is often a concern that could attain its present status only by usurpation.
>
> (Simmel, 1978, p. 482)

The same kind of concerns about this new 'mobile swirling landscape' (Prendergast, 1992, p. 5) can be found in Benjamin (who on particular issues was influenced by Simmel), especially in the by now overly familiar figure of the flaneur, who in Benjamin's work becomes an extinct species but 'only by exploding into a myriad of forms, the phenomenological characteristics of which, no matter how new they may appear, continue to bear his traces, as ur-form. This is the "truth" of the Flaneur, more visible in his after-life than in his flourishing' (Buck Morss, 1989, p. 346). Again the links with mobility are clear;

> Around 1840 it was elegant to take turtles for a walk in the arcades (this gives a conception of the tempo of Flanerie). . . . By Benjamin's time taking turtles for urban strolls had become enormously dangerous for turtles, and only somewhat less so for Flaneurs. The speed-up principles of mass production had spilled over into the streets 'waging war on Flanerie'. *Le Temps* reported in 1936 'Now it is a torrent where you are tossed, jolted, thrown back, carried right and left'. With motor transportation still at an elementary-stage of evolution, one risked being lost in the sea. . . . The utopian moment of Flaneurie was fleeting. But if the Flaneur has disappeared as a specific figure, the perceptive attitude that he embodied saturates modern existence, specifically, the society of mass consumption. In the Flaneur, we recognise our own commonest mode of being in the world.
>
> Benjamin wrote 'the department store is (the Flaneur's) last haunt'. But Flaneur as a form of perception is preserved in the characteristic fungibility of people and things in mass society, and in the merely originary sketch provided by advertising, illustrated journals, fashion and sex magazines, all of which go by the Flaneur principle of 'look, but don't touch'. If mass newspapers demanded an urban readership, more current forms of mass media loosen the Flaneur's essential connection to the city. It was Adorno who pointed to the station-switching behaviour of the radio listener as a kind of aural Flaneurie. In our time, television provides it in optical, non-ambulatory form. In the United States, particularly, the format of news programmes approaches the distracted, impressionistic, physiognomic viewing of the Flaneur, as the sights purveyed take one around the world. And, in connection with world travel, the mass tourist industry now sells Flaneurie in two and four week packets.
>
> (Buck Morss, 1989, pp. 344–6)

But three more recent authors suffice to make the case about the primacy of mobility. First, there is Virilio's work on speed and the emptiness of the quick

(Virilio, 1975, 1977, 1980, 1983, 1988, 1989, 1990, 1991a). So far as Virilio is concerned we live in the 'age of the accelerator' in which 'power is invested in acceleration itself'. For Virilio, speed has many effects. It consumes subjectivity, leading to 'the disappearance of consciousness as a direct perception of the phenomena that informs us about our existence' (Virilio, 1980, p. 16); it also alters the nature of time and narrative, making 'everyone a passer-by, an alien or a missing person' (Virilio, 1989, p. 28); it even drives place out of space so that 'speed is a non-place and the users of transit-spaces, transit-towns (like airports), are spectral-tenants for a few hours instead of years, their fleeting presence is in proportion to their unreality and to that of the speed of their voyage' (cited in Morris, 1987, p. 6). Second, there is Deleuze's work on nomadism, as something that happens between, only temporarily occupying a space and imposing no fixed and sedentary boundaries (Deleuze and Guattari, 1988; Bogue, 1989; Massumi, 1992a; Hardt, 1993). 'The nomadic subject traverses points of pure intensity in migratory fashion. Nomadic thought is a distribution of singular points of possible actualisation/individuation/conductivity; "Nomad thought" does not immure itself in the edifice of an ordered interiority; it moves freely in an element of exterioriority. It does not suppose an identity; it notes difference. It does not respect the artificial division between the three domains of representation, subject, concept and being; it replaces restrictive analogy with a conductivity that knows no bounds' (Massumi, 1988, p. xii). Finally, Schivelsbuch (1986) provides an account of an aesthetics of disappearance, of the landscape as a blur, a streak viewed from a moving platform, 'no longer experienced intensively, discretely, but evanescently, impressionistically – panoramically, in fact' (Schivelsbuch, 1986, p. 189).

(3) This emphasis on ceaseless mobility can be coupled with another related characteristic of the contemporary world, its lack of presence, its *indirectness*. As Derrida and others have been concerned to show, indirect communication,[19] symbolised by the telephone, the postcard and various other ways of communicating at a distance, is a fundamental constant of contemporary everyday life. As Wood (1990) has pointed out, this view also has a long philosophical history in the work of Kierkegaard, Adorno and Heidegger (again), as a study of the relationship between subjectivity and linguistic expression. This emphasis on indirect, elliptical communication extends, of course, into the visual realm. Contemporary ways of seeing are indirect, formed *via* the television and the camera, providing an 'imposture of immediacy' (Virilio, 1989).

(4) The emphasis on ceaseless mobility and indirectness produces a quandary. How can categories, thought of as fixed in space and time with fixed qualities associated with them, be understood in a dynamic, deterritorialised

[19] Derrida and Deleuze and Guattari are perhaps the chief exponents of this emphasis on indirect communication. For example, for Deleuze and Guattari, 'human language, rather than commencing with tropes of direct discourse, begins with indirect discourse, the repeating of someone else's words' (Bogue, 1989, p. 137).

world which admits of no stable entities and so can only be understood in terms of difference, not identity? Three categories cause particular problems: the subject, the body and place. Certainly in a world of mobility, the *subject* is a problematic category. If subjects are nomadic, fractured, heterogeneous and indirectly connected then how can they be understood as agents at all? The answer would appear to be as nomadic points, constantly shifting through strings of subject positions/contexts, in some of which it is possible to speak, in others not, according to prevailing regimes of power (Grossberg, 1992).

> The nomadic subject exists within its nomadic wandering through the ever-changing places and spaces, vectors and apparatuses of everyday life . . . coherent subjectivity is always possible, even necessary, and always effective, even if it is also always fleeting. The subject's shape and effectivity are never guaranteed; its agency depends in part on where it is located, how it occupies its places within specific apparatuses, and how it moves within and between them. . . . The nomadic subject always moves along different vectors, always changing its shape. But it always has an effective shape as a result of its struggle to win a temporary space for itself within the places that have been prepared for it. Nomadic subjects are like 'commuters' moving between different sites of daily life, who are always mobile but for whom the particular mobilities or stabilities are never guaranteed.
>
> (Grossberg, 1988, p. 384)

(5) The problem of boundedness is also important in discussing the body. As Lyotard (1992, p. 50) puts it, 'what is a body . . . in tele-graphic culture?' Bodies have been extended in numerous ways by the new technologies. Equally, they have been intruded into and upon by developments like genetic engineering, new reproductive technologies and plastic surgery (Shilling, 1993; Synott, 1993). Parts of bodies can even be added or subtracted. Yet mobility is not just concerned with the extension of, or intrusion into, the body. It also concerns two other things. First, the history of speed, light and power has led to greater awareness of the mimetic qualities of the body, especially because of the ability of modern mimetic machines to make copies which in turn influence what is copied (Taussig, 1993).[20] This produces a new 'violence of perception' which operates directly on the body. Second, it is important to remember the sheer *joy* of bodies in movement. Movement is a sensuous pleasure in itself. It might even be possible to speculate that the increase in mobility has highlighted and underlined this joy.

(6) The same problem of understanding boundedness also crops up in the case of place. What is *place* in this new 'in-between' world? The short answer is – compromised: permanently in a state of enunciation, between addresses, always deferred. Places are 'stages of intensity', traces of movement, speed and circulation. One might read this depiction of 'almost places' (Kolb, 1990) in Baudrillardean terms as a world of third-order simulacra, where encroaching

[20] Through 'the unstoppable merging of the object of perception with the body of the perceiver and not just with the mind's eye' (Taussig, 1993, p. 25).

pseudo-places have finally advanced to eliminate places altogether. Or one might record places, Virilio-like, as strategic installations, fixed addresses that capture traffic. Or, finally, one might read them, as Morris (1989) does, as frames for varying practices of space, time and speed. There is, in other words, 'no stability in the stopping place' (Game, 1991, p. 166). No configuration of space-time can be seen as bounded. Each is constantly compromised by the fact that what is outside can also be inside.

This problem of 'almost places' (Kolb, 1990) becomes particularly acute in the case of contemporary cities (Shapiro, 1992). It has become increasingly difficult to imagine cities as bounded space-times with definite surroundings, wheres and elsewheres. In Virilio's (1989) terms, they have 'exploded'. As Robins (1991) puts it;

> We can now talk of a process of globalization or transnationalization in the transformation of urban space or form. Manuel Castells describes the advent of the 'informational city', and identifies 'the historical emergence of the space of flows, superseding the meaning of places'. Others have described the same process in similar ways suggesting that 'things are not defined by their physical boundaries any more'. In the place of a discrete boundary in space, demarcating distinct space, one sees spaces co-joined by semi-permeable membranes, exposed to flows of information in particular.
>
> (Robins, 1991, p. 4)

Nowhere does this difficulty of imagining boundedness become clearer than in the increasing mobility and fluidity of the cultures of urban centres, an urban swirl of 'uprooted juxtapositions' (Gitlin, 1989, p. 104), sometimes expressed in totally inadequate phrases like, 'multiculturalism' (Hannerz, 1992).

(7) This same problem of understanding the boundedness of place can be phrased in a more general way: 'all places have borders, even in their centres, and the deconstructive task is to find these borders and the ways in which our constructions cross them while denying that they do' (Kolb, 1990, p. 157). This problem of *borders* has been posed in a number of inter-related ways. It can be worked through as the interaction between the 'global' and the 'local' (whatever these words now mean) (*Theory, Culture and Society*, 1990; Hall, 1991a, 1991b; Hannerz, 1992; King, 1991; Sklair, 1991). For example, in the realm of culture, 'new communications technologies are mobilized in the (re)creation and maintenance of traditions, cultural and ethnic identities which transcend any easy equation of geography, place and culture, creating symbolic networks throughout various . . . communities' (Morley, 1991, p. 14). It can also be worked through as the increasing primacy of indirect communications and relationships, and the consequent blurring of *gemeinschaft* and *gesellschaft* (Anderson, 1983; Calhoun, 1991; Poster, 1990; Wood, 1990). Again, it can be worked through as a questioning of the commonly constructed distinction between a world of intimate lived experience and large-scale social systems (Latour, 1993): 'But *is* there really this distinction? Is it not more correct to say that people increasingly have their world of lived experience

within corporate actors? Does not an expanding share of face-to-face relations arise, unfold and develop *within* organisations?' (Hernes, 1991, p. 124). Last, but not least, it can be worked through as a questioning of the distinction between the public and the private as separate spheres of existence (Rose, 1993).

(8) Perhaps this very indeterminacy of boundaries has led to a new form of institutionalising moment in which systems of domination are based on forms of dispersed power, rather than systematic and generalised forms of repression. This thought, prefigured in the work of the situationists, Foucault, Guattari and others (Plant, 1992) is crystallised by Deleuze (1991) who suggests that we can now see the advent of 'societies of control'. Deleuze distinguishes between three different kinds of societies. The first is societies of sovereignty whose goals and functions were to tax rather than organise production and to rule on death, rather than to administer life. Then came disciplinary societies, which reached their height in the early twentieth century. They instituted vast spaces of enclosure – prisons, hospitals, schools, houses, asylums, factories, that moulded people. Now, new forces knock on the door, societies of control in which ultra-rapid forms of free-floating strategies of control modulate people:

> The numerical language of control is made of codes that mark access to information, or reject it. We no longer find ourselves dealing with the mass/individual pair. Individuals have become 'dividuals', and masses, samples, data, markets or banks. Perhaps it is money that expresses the distinction best since discipline always referred back to minted money that uses gold as a numerical standard, while control related to floating rates of exchange, modulated according to a rate established by a set of standard currencies. The old monetary mole is the animal of the spaces of enclosure, but the serpent is that of societies of control. We have passed from one animal to the other, from the mole to the serpent, in the system in which we live, but also in our manner of living and in our relation with others. The disciplinary man was a discontinuous producer of energy, but the man of control is undulatory, in orbit, in a continuous network (Deleuze, 1991, pp. 5–6). Thus, 'what counts is not the barrier but the computer that tracks each person's position – licit or illicit – and effects a universal modulation' (Deleuze, 1991, p. 7).

(9) The above musings might seem like another Deleuzian dark fantasy of acceleration,[21] moving ever faster both backwards and forwards between a Futurist or Vorticist manifesto and a cyberpunk novel (McCafferty, 1991). But the point is that all is not lost. In societies based on *strategies* of control the 'tactical' everyday practices of 'consumers' may take on a new potency. *Tactics* are, after all, *mobile* ways of using imposed systems: 'immersible and infinitesimal transformations'; 'swarming activity', jostlings for position';

[21] Interestingly, Deleuze writes approvingly of Hume 'As in science fiction, one has the impression of a fictive, strange science-fiction world, seen by other creatures, but also the presentiment that this world is already ours, and these other creatures, ourselves' (Cited in Bogue, 1989, p. 178, Fn. 17).

'errant trajectories' drawing partly unreadable paths across space (de Certeau, 1984, p. xviii). They are pure processes without textual form or realisation: 'doings' (*arts de faire*) like speaking, walking, reading, poaching and tricking (de Certeau, 1984, 1986, 1988, 1991; Frow, 1992; Giard, 1991) which insinuate themselves 'into the other's place, fragmentarily, without taking it over in its entirety, without being able to keep it at a distance' (de Certeau, 1984, p. xix). They are 'challenging mobility that does not respect places', through a 'delinquent narrativity' made up of spatial stories/practices (de Certeau, 1984, p. 130).[22] 'In other words, they are practices through which people can escape without leaving' (Frow, 1992, p. 57). But, 'in our societies, as local stabilities break down, it is as if, no longer fixed by a civilized community, tactics wander out of orbit, making consumers into immigrants in a system too vast to be their own, too tightly woven for them to escape from it' (de Certeau, 1984, p. xx). In other words, mobility now has to challenge mobility.

There are some obvious objections to this developing ontology. I want to list four of them, as well as some possible answers. The first objection is that this ontology can easily be hijacked. Bataille might have reduced it to a carnival of ecstasy and loss. Baudrillard might want to drive to the astral extreme, to a new phase of excess and exorbitance. Harvey might want to write about the effects of the speed-up of circulation time. Jameson might want to write about the Kantian sublime. Derrida might want to write of distant voices and telegramaphones. Lyotard might want to write of telegraphy. And so on. The point is that these are all aspects of this ontology, but they can never be a whole story. In the new elliptical ontology, there is no be-all or end-all. There are only lines of flight.

A second objection is that this sketch of movement, displacement and rapidity, for all its commitment to the 'other', still bears too many traces of eurocentrism, ethnocentrism and androcentrism. It is a rich white heterosexual man's ontology, more akin to tourism than travel; the 'suggestion of free and equal mobility is itself a deception, since we don't all have the same access to the road' (Woolf, 1993, p. 253; see also Pratt, 1992). What about the four-fifths of the world's population that live on one-quarter of its assimilative capacity (one-fifth of which 'live in an unimagineable spiral of despair that the wealthy fifth would not tolerate for 10 seconds' (O'Riordan, 1992, p. 25)) who cannot get up to speed because of lack of income, immigration laws and so on? Or as Kolb (1990, p. 159) puts it, 'where will the septic system go?' Again, what about the ways in which mobility has been gendered? 'Western ideas about travel and the concomitant corpus of voyage literature have generally – if not characteristically – transmitted, inculcated, and reinforced patriarchal values and ideology' (van den Abbeele, 1992, pp. xxv–xxvi). The objection is not that dispossessed do not travel. They clearly do. It is rather that mobility is 'still the wrong language' to describe such activity because it implies that we

[22] Notice that de Certeau does not deny the importance of narrative. Like Carr (1986) he can see no alternative.

are all 'on the road "together"' (Woolf, 1993, p. 235). 'The fantasy of escape from human locatedness' (Bordo, 1990, p. 142) is indulged, even cosseted. This is clearly a charge with some force, but only some. Certainly it is the case that the ceasless migrations of contract workers, the waves of immigrants and refugees, and many of the multitude of women travel in different, more constrained ways, to a different purpose. But it is not immediately clear that the developing ontology of mobility has to be appropriated in a conservative way which would debar these kinds of experiences of what is still a rather mobile 'immobility' (Radway, 1988; Grossberg, 1988). Further, it might also be argued that this ontology of mobility has provided some of the most useful ways of trying to decolonise western ways of thinking by showing that margins and centres, others and selves, exiles and metropoles, are always interconnected. Last, this kind of ontology simply does not lay claim to working out 'the one solution' in quite the way of previous ontologies. It,

> is not a theory which reveals some hidden level of forces, waiting to be enlisted on our behalf. It can free us for more creative gestures and resistances, but it does not by itself take a stand on the issues of the day. To think it does is to change it into a haughty irony and hidden totality. (Godzich, 1986, p. 6)

A third, and perhaps most telling, objection is that this ontology is not as radical as it at first may appear. It simply substitutes one kind of additive infinity, that of the disciplinary society with its logocentric idea that things have finite characteristics and that the world is therefore full of individuated and countable things, for another kind of additive infinity, that of the society of control, where there is 'a grammatological understanding of recurring operations: a constant substitution of functions such that terms simultaneously express and displace previous terms' (Strathern, 1992b, p. 8). Thus a Euro-American frame of meaning is still reproduced. It is this criticism that leaves me at something of a loss, most particularly because if the Euro-American frame of meaning is this pervasive, then it is difficult to know how we can ever move outside of it.

To summarise so far, it seems that an ontology of mobility can challenge eurocentrism, ethnocentrism and androcentrism in three main ways. One is that we are forced to go beyond accounts which insist on regarding the importance of place and placement as self-evident in modern everyday life, as though somehow a mere statement of recognition is enough to provide an anchor, a certitude. Thus Morris (1989) can lay down the following challenge;

> The problem of feminism might be summed up by Prato and Trivero's claim in 'The spectacle of travel' that transport ceased to be a metaphor of Progress when mobility came to characterize everyday life more than the image of 'home and family'. Transport became 'the primary activity of everyday existence'. Feminism has no need whatsoever to claim the home and family as its special preserve, but it does imply a certain discretion about proclaiming its present marginalisation. . . . Yet the sort of claim made by Prato and Trivero does not seek its grounding in historical truth – even the truth of approximation – and this makes

> feminist criticism more difficult. It is meant, perhaps, to be a billboard, a marker in a certain landscape. It marks a recognisable trajectory along which it becomes possible not only for some to think their lives as a trip on a road to nowhere (etc), but for others to think home-and-family as a comfortable, 'empowering' vehicle. . . . So rather than retreating to the invidious position of trying to contradict a bill-board, feminist criticism might make it its own.
>
> (Morris, 1989, p. 68)

Another gain is that an ontology of mobility forces us to think about borders, processes of bordering, and border pedagogy (Tomas, 1992; Giroux, 1992). It makes us consider margins and centres, convergences and overlaps, exiles and evicts (Hebdige, 1993). It insists that we pay more attention to postcolonial, third 'contact' spaces, the spaces in-between which can provide some kind of ground for meeting and dialogue – the creation of 'a speaking and signifying space large enough to accommodate difference, entertain rapprochement' (Carter, 1992, p. 147). For example, it provides us with a way to escape from a growing 'ethnic absolutism' by concentrating on 'fractal patterns of cultural and political exchange that we try to specify through manifestly inadequate theoretical terms like creolisation and syncretism' (Gilroy, 1992, p. 193). The middle, the mediator, and the hybrid are all ok.

Yet another gain is the way in which subjectivity necessarily has to be retheorised in terms of subject positions, strung out in space-time (Spivak, 1987, 1990). In place of a preoccupation with identity as a kind of imploded dialogue, we have to retheorise identity as a space-time distribution of hybrid subject-contexts constantly being copied, constantly being revised, constantly being sentenced and enunciated (Bhabha, 1986; Hall, 1991a, 1991b; Thrift, 1992; Wagner, 1991).

A fourth and final objection is that this ontology can never be made into a detailed intellectual and practical project. That is not the case, I think. The cyborg world of mobility which this ontology is intended to describe is certainly difficult to represent in older perspectivalist terms (Jay, 1992), due in part to the dislocation of absolute time and space. But, in turn, the difficulty of getting a perspective on a mobile world has produced a questioning of orthodox modes of representation like science and narrative, at least as they have been conventionally understood. Science is in question because it is now revealed as a culture without any hard and fast protocols. Narrative is in question because it has too often been subsumed to the one true story, whether directly or by implication (Doel, 1993).

The result is a move towards a new horizon of intelligibility which is perhaps best grasped in the work of de Certeau on what he calls 'science/fiction'. This is a modest attempt to constitute a 'logic' (or heterology) which gives the Other a part and recognises that there is a constant tension between it and its overt representation in discourse. It is therefore a kind of journey (de Certeau talks constantly of voyages, journeys, travelling) in which the 'walk reveals the goddess' (Giard, 1991, p. 219). It is a kind of constant path-clearing. In using the metaphor of journeying/travelling so frequently, de

Certeau is trying to point to three chief things. First, that there is a 'form of truth that is totally alien to me, that I do not discover within myself but that calls on me from beyond me, and . . . requires me to leave the realms of the human and of the sane in order to settle in a land that is under its rule' (Godzich, 1986, p. xvi). Second, to a sense of intellectual itinerary that comes with a recognition of the many forked paths and diversions reading and writing can take. Third, to a mobile sense of 'seeing': 'to travel is to see, but seeing is already travelling' (de Certeau, 1983, p. 26). Thus science/fiction is a third new space of reading/writing between science and fiction, involving elements of both: 'a mixture of narration and scientific practices' dependent upon the metaphor of travelling.[23].

In turn, this emphasis on new regimes of reading and writing based on journeying/travelling leads to new methods of writing based on journeys and travelling. These are hardly a new innovation, of course. Sieburth has pointed to the way in which Sterne consistently experimented with new modes of writing based in travel. What is most singular now is the attempt to write movement in a whole host of ways which involve, to adopt Marcus's (1992) list, both remaking the observed, through problematising the temporal and problematising perspective/voice, and revealing the observer, through making the text's conceptual apparatus explicit, through bifocality, and through critical juxtaposition. Each and every one of these devices involves the use of many sites, both literally and metaphorically.

4. AN ENTIRE FUTURE GEOMETRY

In this section, I want to argue that this new structure of feeling is not just manifested in the higher realms of poststructuralist or postmodernist or postcolonialist theory, it is also to be found in intellectual work of a more practical, empirical bent. I will therefore review in some detail work which seems to me to have been influenced by this structure of feeling. This is not difficult to do. Research in the social sciences and humanities, especially but not only in cultural studies, has become saturated with the vocabulary of mobility – from nomadic criticism and travelling theory through ideas of the ethnographer-as-tourist to the increasing use of metaphors based on maps, topography, billboards, networks, circuits, flows . . . the list goes on. In particular, I will draw on work of a historical bent. This last may seem a strange choice but I would suggest that it is historical work that is particularly likely to show the impact of the new structure of feeling. This is because I take

[23] The idea of a science/fiction can be used in another way as well, to signify the construction of alternative theoretical worlds, rather in the manner of Deleuze:

> he constructs imaginary worlds or alternative universes in the manner of a Borges or a le Guin, showing what reality would be like if it were made up of simulacra, virtual singularities and anonymous forces, or formless bodies and incorporeal surfaces. He invents paradoxical concepts . . . but rather than reinscribe these concepts within traditional texts, he uses them as the building blocks of an alternative world
>
> (Bogue, 1989, p. 159)

it that, to a degree, history tends to project the concerns of the present on to the past, or as Benjamin puts it

> for the materialist historian, every epoch with which he occupies himself is only a fore-history of that which really concerns him. And that is precisely why the appearance of repetition doesn't exist for him in history, because the moments in the course of history which matter most to him become moments of the present through their index as 'fore-history', and change their characteristics according to the catastrophic or triumphant determination of that project.
>
> (Benjamin, 1979, p. 28)

Thus, I think it is possible to see the past being rewritten, from the perspective of our cyborg culture, both in the terms of mobility and as a history of mobility. Indeed so great is this rewriting that mobility is now coming to be seen as the normal state of things, and settlement as the problematic 'other' category that needs to be explored.

Certainly, there is currently a massive outpouring of work concerned with the subject of mobility and the mobility of the subject and I cannot review it all. Rather I will simply make an indicative reference to certain areas of work under four main headings: speed, light, power and writing mobility.

4.1 Speed

The machinic complex of speed includes a number of different but interconnected forms of work. There is quite clearly a growing archive on the subject of travel and travellers stimulated in part by a general realisation that there have been few societies in which travel, contact and mobility have not been important and many more societies, from colonial empires (Innis, 1950; Christopher, 1991) to multinational corporations, in which travel, contact and mobility have been central. They have, in effect, drawn the diagrams of power. There is, first of all, work which considers travellers themselves down through history, in all their diversity. Clifford (1992) provides a bewildering list of missionaries, informants, mixed-bloods, translators, government officers, police, merchants, prospectors, tourists, ethnographers, pilgrims, servants, entertainers, migrant labourers, and immigrants (to which, no doubt, could be added many other travelling figures; explorers, commuters, hobos and tramps, packmen, and so on). The underlying rationale behind this work is that these nomads can produce 'new spaces' which, although they are transient, can sometimes offer new possibilities. An important part of the study of travellers through history has concerned the study of women travellers. It is now realised that women were more mobile than has commonly been realised but mobile in quite different ways (Pratt, 1991; Mills, 1991; Blunt and Rose, 1994).

Second, there is work on travel as a mode of being-in-the world. In particular, there is now more and more work using the Flaneur as a basic category to interrogate the way in which a new ambulatory observer was formed in the nineteenth century city through the 'convergence of new urban space, technologies and symbolic functions of images and products – forms of artificial lighting, new use of mirrors, glass and steel architecture, rail roads, museums, gardens, photographs, fashion crowds' (Crary, 1990, p. 20). Theoretical development and historical research, going hand in hand, have now shown that the category of Flaneur generally is far too gender and class-specific. Yet, at the same time, it contains some more general elements of a particular consumerist mode of being-in-the world (Buck Morss, 1989; Corfield, 1990; Wilson, 1992) which have probably been most clearly developed in studies of the history (see Ousby, 1991) and the current practices of tourists. This work which dates, in particular, from the influential book by MacCannell (1976) on the post-tourist is chiefly concerned with different kinds of traveller and travel experience. More generally it is concerned with different kinds of being in a mobile world and the different kinds of displaced self-understanding that result, chiefly in support of the argument that different kinds of tourist travel valorise the world in different ways (Urry, 1990; Game, 1991). For example, MacCannell (1992) argues that his work on tourism suggests that

> the current dialectic between the global versus local, or sedentary versus nomadic or any other dialectic that involves a contradiction between different levels of socio-cultural organisation, is about to be superseded. The emerging dialectic is between two ways of being out-of-place. One pole is a new synthetic arrangement of life which releases human creativity. The other is a new form of authority, containment of creativity, and control.
>
> (MacCannell, 1992, p. 5)

Third, there is work on the whole range of spatial practices that produce knowledge on and of mobility – maps, diaries, directories, guides, stories, poems, books, magazines, music, photographs, television programmes, videos and the like. In particular, the genre of travel writing – especially travel writing by women – is now coming under scrutiny as a moment in the construction of a particular kind of spatiality/narrativity (Domosh, 1991; Pratt, 1991; Mills, 1991).

Fourth, there is a growing body of work (and especially ethnography) on the not quite/almost places of mobility – roads and railway lines, service stations and airports, hotels and motels, car parks and carports, shopping malls and heritage sites as well as the media of mobility, like the automobile or the aircraft (e.g. Morris, 1987). This is not to forget the extraordinary (and extraordinarily often forgotten) importance of the streets and street life (for example, Corfield, 1990; Glennie and Thrift, 1994).

Fifth, there is now a yearly outpouring of work on the origins and meanings of electronic spaces. For example there is work on the communication spaces of the telephone, radio and fax. There is also work on the visual spaces of the cinema, television and video. More recently, a burgeoning amount of work has been appearing on the cyberspace of computers and computer networks and the new communicational techniques that have been made possible (e.g. Poster, 1990; Lea, 1992).

Sixth, there is the work on borders, frontiers, crossing points and crossroads to be found in the writings of Trinh T. Minh Ha (1989), Rosaldo (1988), Flores and Yudice (1990) and others, from whence comes a realisation of the importance of the way in which spaces, which may well have been originally constructed as places of domination, in which difference and conflict were constructed and lived, also have the potential to become meeting places which promote different kinds of ethical spaces (perhaps this is part of what Derrida means when he writes of opening up new spaces of reading (Wood, 1990)), 'not quite, not white' spaces of 'hybridity and struggle, policing and transgression' (Clifford, 1992, p. 108). The importance of gender and race as transgressive categories becomes clear in these 'contact zones' (Pratt, 1991). More generally, the emphasis on such meeting spaces makes it possible to talk about the conditions under which communication can take place in-between (Price, 1990; Carter, 1992; Karras and McNeil, 1992; Gilroy, 1993).

The current work on speed is perhaps best summarised by two papers. One is a study of the new nomads that transverse the United States (McHugh and Mings, 1991). Each year, hundreds of thousands of Americans and Canadians drive southward toward the sunshine in recreational vehicles, circulating around amongst a network of sites – campgrounds, national parks, resorts and the like. These travellers are usually older, relatively well-off people. Opposed to the structured mobility of these groups of nomads is Hebdige's witness to the travails of the homeless in New York City seen through the lens of Wodickzo's Poliscar, a vehicle for the homeless which is 'a strategy for survival for urban nomads – evicts – in the existing economy' (Wodiczko and Lurie 1988, cited in Hebdige, 1993, p. 181). In one paper, the mobility of the centre, in the other the mobility of the margins, both acting to document and image the social structuring of mobility.

Finally, there is the extensive body of work on immigrant cultures. This last has, of course, existed for a long time. What is new are two things. To begin with, there is the almost/not quite normality of the migrant condition:

> This period of modernity has been characterised by the massive displacement of populations. We are almost all migrants; and even if we have tried to stay at home, the conditions of life have changed so utterly in this century that we find ourselves strangers in our own house. The true novelty is to live in an old country. But despite the normality of displacement, we find the migrant vilified. For alongside the fact of ethnic integration, we also witness a recrudescent nationalism, a yearning for the purification of racial roots and the extermination of alien elements. . . . In this situation it becomes more than ever urgent to develop a framework of thinking that makes the migrant central, not auxiliary to historical processes. We need to disarm the genealogical rhetoric of blood, property and frontiers and to substitute for it a lateral account of local relations, one that stresses the contingency of all definitions of self and the other, and the necessity always to tread lightly. Living in a new country is not an eccentricity; it is the contemporary condition.
>
> (Carter, 1992, p. 3)

Following on from this first point, migrant cultures are no longer seen as copies from an original, but as in constant interaction with 'home' cultures and, at the same time, evolving in their own ways, in part through contact with other cultures, often influencing the practices of their 'home' cultures as they do so (Ganguly, 1992). These cultures are now seen as 'stratified' processes, their origins in question (Friedman, 1992), their destinations in suspension. They are crossroads at crossroads (James, 1992).

4.2 Light

Here I want to point chiefly to the growing literature on vision and visuality (Foster, 1988). Much of this work has been stimulated by just the serial, reproducible image of still and cinema photography. In particular, there is now a large body of historical work on photography and the cinema and television. From this work, it is clear that photography, the cinema and television have to be seen in at least three ways. First, they are an archive, albeit one that can be interpreted in many ways. As Porter (1991b, p. 211) notes, in a discussion on the body and photography,

> we possess a photographic record now stretching back almost a century and a half of people's physical appearances. Once again, there is no need to belabour the misinterpretations which would result from a naive reliance on the veracity of visual images; of course, the camera lies or, more precisely, photographs are not snapshots of reality but, like paintings, form cultural artefacts conveying complicated coded conventional signs to primed 'recorders'.
>
> But this caveat applies to some photographs more than to others. Posed portraits capture how people wish to be remembered, all scrubbed and dolled up in their Sunday best. But Victorian photographers were also fond of taking 'documentary' street snapshots and these caught people in their everyday moments, gestures, and as a result recorded such aspects as body language and social space more informatively than any printed text. The photographic archive reveals and confirms a great deal about both the physical transformations of the human condition in modern times (ageing, deformities, malnutrition, etc) and what Goffman has called 'the presentation of self' (body language, gestures, the appropriation of physical self). Photographs remain oddly underexploited as a historical resource.
>
> (Porter, 1991b, p. 211)

Second, they are also a means of perception. They tell us about different modes of visualisation according to which primary social processes and institutions function (Crary, 1990). For example, according to some commentators, we now live, or have lived until recently, in a society based on the semiotics of the spectacle (Debord, 1966). Whether this is the case or not, the idea has proved to have some currency in recent historical work as a model of certain modes of visualisation which makes it possible to locate developments in the past as in, for example, Richards' (1990) examination of late Victorian advertising or Crary's (1990) study of late eighteenth and early nineteenth century techniques of observation. In turn, the system of spectacle can be seen to be historically transient;

> Recently, Baudrillard has raised the possibility that capitalism has itself brought about 'the very abolition of the spectacular', by which he means that current technologies have far surpassed the spectacular system of representation. . . . Selling things has now become the domain of cybernetics, the modern science concerned with analysing the flow of information in electronic, mechanical pulses, blips on a television screen, and patented genetic structures count as quantifiable commodities, the days of spectacle may well be numbered. In the years to come it may turn out that the semiotics of spectacle played a transitional role in capitalist mythology. . . . For now it makes sense to render spectacle as one element among many in modern commodity culture, one manifestation of what Martin Jay has called the many 'scopic regimes of modernity'.
>
> (Richards, 1990, p. 258)

Third, they are a means of framing and constituting the body, and especially the body in motion. Thus Featherstone (1991, p. 178) has pointed to the way in which photography and film helped to create new standards of appearance and bodily presentation, at first chiefly for women but subsequently for men too, which would 'enable the body to pass muster under the camera'.

4.3 Power

Here I want to indicate the growing interest in electrical power, especially as a root metaphor in our societies, linked both to light (and the image) and to speed (and communication). This literature takes on three main forms. The first of these is quite straightforward. It is the growing literature on the social and cultural history of electrical power and the way in which

> in daily experience, adopting electricity changed the appearance and multiplied the meanings of the landscapes of life, making possible the street car suburb, the Department store, the amusement park, the assembly-line factory, the electrified home, the modernised farm, and the utopian extension of all these, the world's fair.
>
> (Nye, 1990, p. x)

Such histories, which point to major differences in the ways in which electrical power was adopted and interpreted in different places, have now begun to burgeon (e.g. Nasaw, 1992; Nye, 1990; Luckin, 1990; Marvin, 1988; Anderson, 1991).

The second and third forms both concern the nexus between electricity and the concepts of the body, self and identity, and the subject. Thus the second form is the way in which work on the logistics and erotics of the body, self and identity and even the subject are described in terms of different forces and intensities. What was a psychoanalytic model (with interesting links to electricity) has become a general means of representing the social reproduction of desire and affect and even, in the case of Deleuze and Guattari (1988), the 'body' of capital.

The third form concerns the articulation between the body, self and identity, and the subject and the information technologies of different eras. Thus Kittler (1990) suggests that the new information technologies of the turn of the century transformed the relationship between hand, eye and letter in such a way that the transition between the self that writes and the writing of the self was disputed, leading to new forms of writing and arche-writing of the kind that Derrida has tried to interrogate. More dramatically, Seltzer (1992) suggests that these new technologies put the anthropomorphism of writing to rest for all time and put in its place, through 'the violent immediacy promised by communication and control technologies operated by the electric signal or button' . . . a 'pure performative that instantly connects conception, communication and execution'. This pure performative:

> is legible, for instance in the rapid adoption of the electric chair and the 'deadly current' as the socially acceptable form of execution in the 1880s. . . . It is legible also in the communication technology of the telephone that could order or stay the execution. . . . And it is everywhere legible in the links between man-factories, death-factories, electric signals and body counts, and in the small marginal movement of the hand that, in effect, communicates execution.
>
> (Seltzer, 1992, pp. 11–12)

Poster (1990) takes this argument farther again, arguing that the new level of interconnectivity afforded by current electronic media and the success of electronically configured language have produced new forms of the body, self and subject in which 'the body . . . is no longer an effective limit of the subject's position' (Poster, 1990, p. 15) and 'new and unrecognisable forms of communicating are in formation' (Poster, 1990, p. 154).

4.4 Writing mobility

Finally, some note has to be taken of the problems of communicating mobility, of incorporating, to précis the intent of Benjamin's significantly named *One Way Street*, 'the outside world of gas stations, metros, traffic noises, and neon lights . . . into the text' (Buck Morss, 1989, p. 17). These problems are, I think, extreme.

It is undoubtedly the case that communicating mobility demands a change in style of 'writing'. Indeed, as I have argued, that might be one of the ways in which this structure of feeling might be able to be described. Yet, on the whole, while declarations of intent have abounded, actual demonstrations have been rather thin on the ground. The reasons for this are not always clear. One may be that it is difficult to represent something that is not always meant to be open to representation. Ultimately, liminologies are, well, liminal. Another reason is that writing is not always the most appropriate form of representation of the elliptical world that we now live in (see Gregory, 1991). Communicating mobility may well demand the incorporations of other forms of expression that the written word – all kinds of visual images, laser discs, video, and so on

– but we are not practised at these (Crawford and Turton, 1992).[24] A further reason is that all the possibilities of the new hybrid 'electric text' that meld texts and images in new software packages, hypertext and so on, have still not been fully thought through. Finally, there is the problem of what metaphors to use to mark out a communicative scheme – journeys or roads, successions of places, billboards or landmarks, or even messages like postcards or electronic mail. We may still need narrative but it is clear, given this list of options, that traditional narrative is often ill-suited to communicating mobility.

This may, of course, be one of the reasons why some writers have turned to science for inspiration. This turn has taken two chief forms. There is, first of all, the emphasis on scientific metaphors as ways of thinking/writing mobility as, for example, in the case of Derrida's use of cybernetic and, latterly, biogenetic themes (Johnson, 1993). Then, second, there is the increasing interest in recent science fiction, both because of inherent motifs like technological sentience, hydridity, indeterminacy, transgression, and the like, and because of the examples it provides of how to communicate these motifs (Kuhn, 1990; Bukatman, 1993).

The result of many of the writing experiments currently under way can, like much 'techno-art' (Pomeroy, 1991), be irredeemably awful but given the plethora of worthy but dull academic publications it is difficult to criticise the impulse, if not the result of these arguments. Further, they can work. I think of Meaghan Morris's (1987) 'At Henry Parkes Motel', parts of Paul Carter's *The Road to Botany Bay* (1987), and *Living in a New Country* (1992) as well as more tangential works like Richard Price's *Alabi's World* (1989) or Susan Buck Morss's *The Dialectics of Seeing* (1989) which have tried to write creatively, from many different positions, using a range of techniques to communicate the modern world's 'stereospecificity' (Johnson, 1993).

5. CONCLUSIONS

Clearly a paper which is intending to describe a structure of feeling called mobility should not conclude. It should keep on the move. Just as clearly, academic convention suggests at least a few parting thoughts are in order.

In this chapter, I have tried to show the development of a new structure of feeling that I have called 'mobility'. This structure of feeling is based in a cyborg culture in which the boundaries between humans and machines have become ever more permeable, leading to more and more actant subjects and to an increasing emphasis on machinic metaphors and practices. It seems to me to be important to recognise this state of affairs and to work with it. But care needs to be taken. There are some important caveats that need to be entered concerning this chapter. Two of these seem particularly important.

The first of these is that a cultural hypothesis like mobility can only be driven so far without breaking down. Some things are not as mobile as some

[24] The increasing publication of graphic novels suggests another communicational idea.

writers like to claim; buses and trains are often late; the average speed of automobile traffic in London is less than ten mph; a large percentage of trips in Britain are on foot, and walking has probably never been a more popular recreation. Again, many people in western societies are still hardly mobile at all; elderly people stuck in nursing homes, inner-city children who have never seen the sea though they live but a few miles from it; whilst, for all the population, waiting is an endemic fact of life (see Buck Morss, 1989, p. 104, or Lash and Urry, 1993, on waiting). Then again, mobility requires paths, or roads, or airlines, or digital highways. Mobility is not a free-for-all; it is guided by networks etched on to the ground, by air corridors, by frequencies, by maps, by itineraries, even by parking restrictions. Then, not least, every journey includes stops in places; continuity requires discontinuity (Game, 1991). In turn, these stops may frame interaction at a more pedestrian pace: 'Locomotion. The slower the better. Stopping at question marks' (Sauer, 1956, p. 296). In other words, mobility is a cultural *hypothesis* which involves clear and dangerous elements of exaggeration. On the other hand, it also has to be asked whether the modern world can be understood (or even seen) through a sedentary gaze.

But it is not just about exaggeration. The problem is more deep-seated than that. It is a matter of gender. This is the second caveat. Feminist writers like Woolf have argued that the idea of mobility is strongly gendered and has to be appropriated in different ways before it can be used effectively. I have argued that these writers are partly wrong but, in certain circumstances, they are also partly right. It is difficult not to come to the conclusion that certain of those who have written round mobility, like Deleuze and Guattari 'have uncritically assimilated the modernist ethos of incessant self-transformation, becoming, and psychic instability' (Best and Kellner, 1991, p. 107) and it is difficult not to come to the conclusion that part of the reason for this is concerned with the gendered nature of their writing. As Jardine (1984, p. 59) writes, 'when enacted, when performed (the promises offered by Deleuze and Guattari's theory) are to be kept only between bodies gendered male. There is no room for women's bodies and their other desires in these creatively limited, mono-sexual, brotherly machines'. Issues of intersubjectivity and the social, of actants, are simply not fully addressed.[25]

This is why this chapter has relied so heavily on the work of writers like Haraway, Latour and Strathern. Their writings add another dimension to the structure of feeling I have called mobility and they do this in three ways. First, the are not caught up in the 'heady cartographic fantasy of the powerful' (Ross, 1991, p. 148) that some of the writing on mobility tends to simulate. Neither do they play to the male fantasy of the 'technobody' with its prosthetic add-ons, all meant to boost masculinity, which were such a prevalent theme of the 1980s, 'the enhancements and retrofits . . . that boys always dreamed of

[25] Similar criticisms have been made from another angle, in work on science and nature. See, for example, Jordanova (1986, 1989); Driver and Rose (1992), Rose (1993).

having, but . . . were also body-altering and castrating in ways that boys always had nightmares about' (Ross, 1991, p. 153). In other words, these writings do not worship technology, either by celebrating it or by bewailing it. In so far as 'technology' is seen as oppressive it is as part of the social organisation of domination, as a part of new forms of regulatory authority.

Second, notions like the 'cyborg', the hybrid or the mediator are all intended to take back some of the ground ceded to the technology as salvation, or doom, merchants. The ever more permeable boundaries between humans and machines are seen as not only a product of power relations but also as a site with potential for contesting and reworking those relations. In other words, these writings believe in the power of transgression as well as regulatory authority, especially for women who, traditionally at least, are meant to be less technoliterate and more likely to be the victims of technology.

Third, these writings are an important part of the contemporary structure of feeling called mobility because they deny the opposition between a vaulting culture of technology and an untainted space of 'Edenic naturalism' (Ross, 1991). The opposition between technology, and society more generally, and nature, with human individuals at the hinge, is found to be groundless. Instead, what there is is a parliament of things which have different degrees and kinds of sentience and which, therefore, take us away from the question of humanism, with its species-centrism, and towards the question of what it is to be 'the human in a post-humanist landscape' (Haraway, 1992a, p. 86) – or what Strathern (1992) calls 'the aesthetics of personification'. In other words, entities become harder and harder to identify as identifiable entities. They are 'variable geometry' entities (Latour, 1993, p. 107). In turn, Latour argues that this state of affairs requires a new constitution which divides up beings, the properties ascribed to them, and their acceptable forms of mobilisation, in different ways. Or as Strathern (1992a, p. 174) argues, we need to debate 'the kinds of things it is conceivable to think about in the late twentieth century'.

In turn, such a point of view quite clearly links to the growing 'leakiness of distinctions' (Haraway, 1985) in other domains as well; not only between the human and the technological world but also the human and the natural world (as well as 'within' the human world). Thus to 'boundary' debates on artificial intelligence, the extension of vision, and prosthetics, need to be added boundary debates on animal rights, in-vitro fertilisation, transplants and genetic engineering. Debates like these have three functions. First, and most simply, they are the processes through which societies make culture explicit – through boundary-setting. Second, they bring into question standard modernist futures since they operate 'merographically' across divides which are normally considered to be connected, even analogous, but also different, not equal, which now do not appear to be as different, or as unequal, as they once did – just as the objects of attention (the technosphere, nature) no longer appear as different as they once did. Thus, 'facts' of the technosphere can become 'facts' of culture and vice versa and 'facts' of nature can become facts of culture and vice versa. In turn, the kind of analogies and metaphors we can draw on no longer act well and others have to be created. Thus, conventional notions of 'the human' are gradually overturned because there is *no permanent representation.* Third, the 'human' can no longer be seen as a hinge between the poles of the technological and the natural. As Latour (1993) puts it:

> If the human does not possess a stable form, it is not formless for all that. If instead of attaching it to one constitutional pole or the other, we move it closer to the middle, it becomes the mediator and even the intersector of the two. The human is not a consitutional pole to be opposed to that of the nonhuman. The two expressions 'humans' and 'nonhumans' are belated results that no longer suffice to designate the other dimension . . . we should be talking about morphism. Morphism is the place where technomorphisms, zoomorphisms, plausimorphisms, ideomorphisms, theomorphisms, sociomorphisms, psychomorphisms, all come together. Their alliances, and their exchanges, taken together, are what define the anthropos. A weaver of morphisms – Isn't that enough of a definition?
>
> (Latour, 1993, p. 137)

To summarise once more, the structure of feeling which I have attempted to outline is one which is an attempt to understand the experience of a culture which is a part of a world which is increasingly manufactured, increasingly far from a 'natural' environment (whatever that might be), a third rather than a second nature in which machines have their place. In the 1960s, Lewis Mumford used the word 'megatechnics' to describe this unprecedented stage of mechanisation;[26]

> In terms of the currently accepted picture of the relation of man to technics, our age is passing from the primeval state of man, marked by the invention of tools and weapons, to a radically different condition, in which he will not only have conquered nature but detached himself completely from the organic habitat. With this new megatechnics, he will create a uniform all-enveloping structure, designed for automatic operation.
>
> (Mumford, 1966, p. 303)

Nearly thirty years later, this kind of utterance seems to be both exaggerated (in the way in which it forecasts a phallocentric dominion over nature) and too cautious (in that it failed to foresee the birth of a new kind of hybrid nature). Yet the impulse to understand a manufactured 'microtechnic' world, and give voice to it, seems just as important. A change of style indeed.

ACKNOWLEDGEMENTS

This chapter was first presented as the plenary address to the Cukanzus conference in Vancouver in August, 1992. Many of the comments received subsequently have proved very useful in redrafting the paper. The redraft of the paper proved possible because of the time and facilities provided by the Netherlands Institute of Advanced Study.

[26] By the 1960s Mumford had entered a more pessimistic phase. Perhaps his earlier more optimistic writings, especially *Technics and Civilization* (1934) are in greater harmony with this paper.

Parts of this paper first appeared in a different form in 'For a New Regional Geography 3' in *Progress in Human Geography*, 1993.

REFERENCES

Adam, B. (1991) *Time and Social Theory*, Polity Press, London.

Adas, M. (1989) *Machines as the Measure of Men. Science, Technology and Ideas of Western Dominance*, Cornell University Press, Ithaca.

Adler, J. (1989) Travel as performed art, *American Journal of Sociology*, Vol. 94, pp. 1366–91.

Anderson, B. (1983) *Imagined Communities*, Verso, London.

Armstrong, T. (1991) The electrification of the body, *Textual Practice*, Vol. 8, pp. 16–32.

Attali, J. (1991) *Noise*, Marion Boyars, Edinburgh.

Augé, M. (1986) *Un Ethnologue dans le Métro*, Hachette, Paris.

Bachelard, G. (1961) *The Poetics of Space*, Northwestern University Press, Evanston.

Baldwin, J. (1955) *Notes of a Native Son*, Beacon Press, Boston.

Balibar, E. (1972) in L. Althusser and E. Balibar, *Reading Capital*, Verso, London.

Barthes, R. (1981) *Camera Lucida. Reflections on Photography*, Noonday, New York.

Batchen, G. (1991) Desiring production itself: notes on the invention of photography, in R. Diprose and R. Ferrell (eds.) *Cartographies*, Allen & Unwin, Sydney, pp. 13–26.

Baudrillard, J. (1988) *America*, Verso, New York.

Beck, U. (1992) *Risk Society. Towards a New Modernity*, Sage, London.

Beninger, J. R. (1986) *The Control Revolution. Technological and Economic Origins of the Information Society*, Harvard University Press, Cambridge, Mass.

Benjamin, W. (1973) *Charles Baudelaire*, Verso, London.

Benjamin, W. (1979) *One Way Street and Other Writings*, New Left Books, London.

Benton, T. (1993) *Natural Relations. Ecology, Animal Rights and Social Justice*, Verso, London.

Best, S. and Kellner, D. (1991) *Postmodern Theory. Critical Interrogations*, Macmillan, London.

Bhabha, H. (1986) Signs taken for wonders: questions of ambivalence and authority under a tree outside Delhi, May, 1817, in H. L. Gates (ed.) *Race, Writing and Difference*, University of Chicago Press.

Bijker, W., Hughes, T. P. and Pinch, T. (eds.) (1989) *The Social Construction of Technological Systems*, MIT Press, Cambridge, Mass.

Bijker, W. and Law, J. (eds.) (1991) *Shaping Technology/Building Society. Studies in Sociotechnical Change*, MIT Press, Cambridge, Mass.

Blunt, A. and Rose, G. (eds.) (1994) Guilford Press, New York.

Bogue, R. (1989) *Deleuze and Guattari*, Routledge, London.

Bordo, S. (1990) Feminism, postmodernism and gender-scepticism, in L. Nicholson (ed.) *Feminism/Postmodernism*, Routledge, London, pp. 133–56.

Bottomley, G. (1992) *Out of Place*, Cambridge University Press.

Bourdieu, P. (1984) *Distinction*, Routledge and Kegan Paul, London.
Bourdieu, P. (1988) *Homo Academicus*, Polity Press, Cambridge.
Bourdieu, P. (1990) *In Other Words. Essays towards a Reflexive Sociology*, Polity Press Cambridge.
Bourdieu, P. (1991) Epilogue. On the possibility of a field of world sociology, in P. Bourdieu and J. S. Coleman (eds.) *Social Theory for a Changing Society*, Westview Press, Boulder, pp. 373–87.
Briggs, A. (1989) *Victorian Things*, Batsford, London.
Briggs, A. (1992) The later Victorian age, in B. Ford (ed.) *Victorian Britain. The Cambridge Cultural History, Volume 7*, Cambridge University Press, pp. 2–38.
Buck Morss, S. (1989) *The Dialectics of Seeing. Walter Benjamin and the Arcades Project*, MIT Press, Cambridge, Mass.
Bukatman, S. (1993) *Terminal Identity. The Virtual Subject in Postmodern Fiction*, Duke University Press, Durham, North Carolina.
Burnett, J. (1986) *A Social History of Housing*, Methuen, London.
Buzard, J. (1993) *The Beaten Track. European Tourism, Literature and the Ways to Culture 1800–1918*, Oxford University Press.
Byatt, I. R. (1979) *The British Electrical Industry 1875–1914*, Oxford University Press, Oxford.
Calhoun, C. (1991) Indirect relationships and imagined communities: large-scale social integration and the transformation of everyday life, in P. Bourdieu and J. S. Coleman (eds.) *Social Theory for a Changing Society*, Westview, Boulder, pp. 95–121.
Carey, J. W. (1989) *Communication as Culture. Essays on Media and Society*, Routledge, London.
Carr, D. (1986) *Time, Narrative and History*, Indiana University Press, Bloomington.
Carter, P. (1987) *The Road to Botany Bay*, Faber, London.
Carter, P. (1992) *Living in a New Country. History, Travelling and Language*, Faber & Faber, London.
Chesnaux, J. (1992) *Brave Modern World*, Thames & Hudson, London.
Clifford, J. (1992) Travelling cultures, in L. Grossberg, C. Nelson and P. Treichler (eds.) *Cultural Studies*, Routledge, London, pp. 96–116.
Coghlan, A. (1992) *New Scientist*, 23rd September.
Cohen, M. (1993) *Profane Illumination. Walter Benjamin and the Paris of Surrealist Revolution*, University of California Press, Berkeley.
Cohen, S. (1986) *Historical Culture: On the Recoding of an American Discipline*, University of California Press, Berkeley.
Collins, H. M. (1990) *Artificial Experts. Social Knowledge and Intelligent Machines*, MIT Press, Cambridge, Mass.
Corfield, P. (1990) Walking the city streets: social role and social identity in the towns of 18th C England, *Journal of Urban History*, Vol. 16, pp. 132–74.
Cowan, R. S. (1983) *More Work for Mother. The Ironies of Household Technology from the Open Heart to the Microwave*, Basic Books, New York.
Crary, J. (1990) *Techniques of the Observer. On Vision and Modernity in the Nineteenth Century*, MIT Press, Cambridge, Mass.

Crawford, D. and Turton, D. (eds.) (1992) *Film as Ethnography*, Manchester University Press.

Cubitt, S. (1991) *Timeshift. On Video Culture*, Routledge, London.

de Certeau, M. (1983) *The Writing of History*, Columbia University Press, New York.

de Certeau, M. (1984) *The Practice of Everyday Life*, University of California Press, Berkeley.

de Certeau, M. (1986) *Heterologies*, Manchester University Press.

de Certeau, M. (1988) *The Writing of History*, Columbia University Press, New York.

de Certeau, M. (1991) Travel narratives of the French to Brazil: sixteenth to eighteenth centuries, *Representations*, Vol. 33, pp. 221–6.

Delanda, M. (1990) *War in the Age of Intelligent Machines*, Zone Books, Cambridge, Mass.

Delanda, M. (1991) *War in the Age of Intelligent Machines*, Zone Books, New York.

Deleuze, G. (1991) *October*.

Deleuze, G. and Guattari, F. (1983) *On the Line*, Semiotext(e), New York.

Deleuze, G. and Guattari, F. (1988) *A Thousand Plateaux. Capitalism and Schizophrenia*, Athlone Press, London.

Der Derian, J. (1992) *Antidiplomacy. Spies, Terror, Speed and War*, Blackwell, Oxford.

De Sola Pool, I. (1977) *The Social History of the Telephone*, MIT Press, Cambridge, Mass.

Doel, M. (1993) Proverbs for paranoids: writing geography on hollowed ground, *Transactions, Institute of British Geographers* (forthcoming).

Domosh, M. (1991) Towards a feminist historiography of geography, *Transactions of the Institute of British Geographers*, N.S. 16, 95–104.

Dreyfus, H. L. (1991) *Being-in-the-World. A Commentary on Heidegger's Being and Time, Division 1*, MIT Press, Cambridge, Mass.

Dreyfus, H. (1992) Heidegger's history of the being of equipment, in H. L. Dreyfus and H. Hall (eds.) *Heidegger. A Critical Reader*, Blackwell, Oxford, pp. 173–85.

Driver, F. and Rose, G. (eds.) (1992) Nature and Science. Essays in the History of Geographical Knowledge, *Historical Geography Research Series No. 28*, London.

Emberley, P. (1989) Places and Stories: the Challenge of Technology, *Social Research*, Vol. 56, pp. 241–85.

Eribon, D. (1991) *Michael Foucault*, Harvard University Press, Cambridge, Mass.

Featherstone, M. (1991) The body in consumer culture, in M. Featherstone, M. Hepworth and B. Turner (eds.) *The Body. Social Process and Cultural Theory*, Sage, London, pp. 170–96.

Feenberg, A. (1991) *Critical Theory of Technology*, Oxford University Press, New York.

Flores, J. and Yudice, G. (1990) Living borders/buscando America, *Social Text*, Vol. 24, pp. 57–84.

Forty, A. (1986) *Objects of Desire*, Thames & Hudson, London.

Foucault, M. (1973) *The Order of Things*, Tavistock, London.

Friedman, J. (1992) Narcissism, roots and postmodernity: the constitution of selfhood in the global crisis, in S. Lash and J. Friedman (eds.) *Modernity and Identity*, Blackwell, Oxford, pp. 331–66.
Frow, J. (1992) Michel de Certeau and the politics of representation, *Cultural Studies*, Vol. 5, pp. 52–60.
Game, A. (1991) *Undoing the Social*, Open University Press, Milton Keynes.
Ganguly, K. (1992) Migrant identities: personal memory and the constitution of selfhood, *Cultural Studies*, Vol. 6, pp. 27–50.
Gellner, E. (1974) The new idealism: course and meaning in the Social Sciences, in A. Giddens (ed.) *Positivism and Sociology*, Heinemann, London.
Giard, L. (1991) Michel de Certeau's heterology and the New World, *Representations*, Vol. 33, pp. 212–21.
Giddens, A. (1991) *Modernity and Self-Identity*, Polity Press, Cambridge.
Gilroy, P. (1992) Cultural studies and ethnic absolutisms, in L. Grossberg, C. Nelson and P. Treichler (eds.) *Cultural Studies*, Routledge, London, pp. 187–98.
Gilroy, P. (1993) *The Black Atlantic. Modernity and Double Consciousness*, Verso, London.
Giroux, H. A. (1992) Revisiting difference: cultural studies and the discourse of critical pedagogy, in L. Grossberg, C. Nelson and P. Treichler (eds.) *Cultural Studies*, Routledge, London, pp. 199–212.
Glennie, P. G. and Thrift, N. J. (1994) Gender and consumption, in N. Wrigley and M. Lowe (eds.) *Retailing. Consumption and Capital*, Longman, London.
Godzich, W. (1986) Foreword: the possibility of knowledge, in M. de Certeau (ed.) *Heterologies. Discourse on the Other*, Manchester University Press, pp. vii–xxi.
Gregory, D. (1991) Interventions in the Historical Geography of Modernity, *Geografiska Annaler*, no. 733, pp. 17–44.
Grossberg, L. (1988) Wandering audiences, nomadic critics, *Cultural Studies*, Vol. 2, pp. 377–91.
Hall, S. (1991a) The local and the global: globalisation and ethnicity, in A. D. King (ed.) *Culture, Globalisation and the World-System*, Macmillan, London, pp. 19–40.
Hall, S. (1991b) Old and new identities, old and new ethnicities, in A. D. King (ed.) *Culture, Globalisation and the World-System*, Macmillan, London, pp. 41–68.
Hannah, L. (1979) *Electricity Before Nationalization*, Methuen, London.
Hannerz, U. (1992) *Cultural Complexity. Studies in the Social Organization of Meaning*, Columbia University Press, New York.
Haraway, D. J. (1985) Manifesto for cyborgs, science, technology, and socialist feminism in the 1980s, *Socialist Review*, Vol. 80, pp. 65–108.
Haraway, D. J. (1991a) *Simians, Cyborgs, and Women. The Reinvention of Nature*, Free Association Books, London.
Haraway, D. J. (1991b) The actors are cyborgs, nature is coyote, and the geography is elsewhere: postscript to cyborgs at large, in C. Penley and A. Ross (eds.) *Technoculture*, University of Minnesota Press, Minneapolis, pp. 21–6.

Haraway, D. (1992a) The promises of monsters: a regenerative politics for imappropriate/d others, in L. Grossberg, C. Nelson and P. Treichler (eds.) *Cultural Studies*, Routledge, New York, pp. 295–337.

Haraway, D. (1992b) *Simians, Cyborgs, and Women: The Reinvention of Nature*, Free Association Books, London.

Haraway, D. J. (1992c) Ecce Home, Aint (Arnt) I a woman, and Inappropriate(d) others: the human in a post-humanist landscape, in J. Butler and J. W. Scott (eds.) *Feminists Theorise the Political*, Routledge, New York, pp. 86–100.

Haraway, D. J. (1993) Modest witness @ second millenium: the female man meets oncomouse. Paper given to British Association of Social Anthropologists, Oxford, July.

Hardt, M. (1993) *Gilles Deleuze. An Apprenticeship in Philosophy*, UCL Press, London.

Hayden, D. (1981) *The Grand Domestic Revolution. A History of Feminist Designs for American Homes, Neighbourhoods and Cities*, MIT Press, Cambridge, Mass.

Hebdige, D. (1993) Redeeming witness: in the tracks of the homeless vehicle project, *Cultural Studies*, Vol. 7, pp. 173–223.

Hernes, G. (1991) Comments in P. Bourdieu and J. S. Coleman (eds.) *Social Theory for a Changing Society*, Westview, Boulder, Colorado, pp. 121–6.

Hughes, T. P. (1983) *Networks of Power: Electrification in Western Society 1880–1930*, Johns Hopkins University Press, Baltimore.

Innis, H. (1950) *Empire and Communications*, Oxford University Press.

James, W. (1992) Migration, racism and identity: the Caribbean experience in Britain, *New Left Review*, no. 193, pp. 15–55.

Jameson, F. (1991) *Postmodernism, or the Cultural Logic of Late Capitalism*, Verso, London.

Jardine, A. (1984) Women in Limbo: Deleuze and his Br(others), *Substance*, Vol. 13, pp. 46–60.

Jay, M. (1992) Scopic regimes of modernity, in S. Lash and J. Friedman (eds.) *Modernity and Identity*, Blackwell, Oxford, pp. 178–95.

Jayaweera, N. D. (1983) Communication satellites: a third world perspective, *Media Development*, 30.

Jenkins, H. (1992) *Textual Poachers. Television and Participating Culture*, Routledge, New York.

Johnson, C. (1993) *System and Writing in the Philosophy of Jacques Derrida*, Cambridge University Press.

Jordanova, L. (ed.) (1986) *Languages of Nature. Critical Essays on Science and Literature*, Free Association Books, London.

Jordanova, L. (1989) *Sexual Visions. Images of Gender in Science and Medicine in the Eighteenth and Twentieth Centuries*, Harvester Wheatsheaf, Brighton.

Karras, A. and McNeil, J. R. (eds.) (1992) *Atlantic American Societies*, Routledge, London.

Kern, S. (1983) *The Culture of Time and Space 1880–1918*, Harvard University Press, Cambridge, Mass.

King, A. D. (ed.) (1991) *Culture, Globalisation and the World-System*, Macmillan, London.

Kittler, F. A. (1990) *Discourse Networks 1800/1900*, Stanford University Press.
Kolb, D. (1990) *Postmodern Sophistications. Philosophy, Architecture and Tradition*, University of Chicago Press.
Kenner, H. (1987) *The Mechanic Muse*, Oxford University Press, New York.
Land, N. (1992) Circuitries, in J. Broadhurst (ed.) *Deleuze and the Transcendental Unconscious*, University of Warwick Press, Coventry.
Lash, S. and Urry, J. (1993) *Economies of Signs and Spaces*, Sage, London.
Lamont, M. (1987) How to become a dominant French Philosopher, *American Sociological Review*, Vol. 93, pp. 584–622.
Latour, B. (1993) *We Have Never Been Modern*, Harvester Wheatsheaf, Brighton.
Lea, M. (ed.) (1992) *Contexts of Computer-Mediated Communication*, Harvester Wheatsheaf, Brighton.
Leed, E. J. (1991) *The Mind of the Traveller. From Gilgamesh to Global Tourism*, Basil Books, New York.
Lefebvre, H. (1991) *The Production of Space*, Blackwell, Oxford.
Lingis, A. (1992) The society of dismembered body parts, in J. Broadhurst (ed.) *Deleuze and the Transcendental Unconscious*, University of Warwick Press, Coventry, pp. 1–20.
Lodge, D. (1991) *Paradise News*, Secker & Warburg, London.
Lucic, S. (1991) *Charles Sheeler*, New York.
Luckin, B. (1990) *Questions of Power. Electricity and Environment in Inter-War Britain*, Manchester University Press.
Lynch, K. (1972) *What Time is This Place*? MIT Press, Cambridge, Mass.
Lyotard, J. F. (1992) *The Inhuman. Reflections on Time*, Stanford University Press.
MacCannell, D. (1976) *The Tourist. A New Theory of the Leisure Class*, Schocken, New York.
MacCannell, D. (1992) *Empty Meeting Grounds. The Tourist Papers*, Routledge, London.
Mack, P. A. (1990) *Viewing the Earth. The Social Constitution of the Landsat Satellite*, MIT Press, Cambridge, Mass.
Major-Poetzel, P. (1979) *Michel Foucault's Archaeology of Western Culture*, University of North Carolina Press, Chapel Hill.
Mandel, E. (1978) *Late Capitalism*, Verso, London.
Marcus, G. (1992) Past, present and emergent identities: requirements for ethnographies of late twentieth century modernity worldwide, in S. Lash and J. Friedman (eds.) *Modernity and Identity*, Blackwell, Oxford, pp. 309–30.
Marx, K. (1856) *Economic and Philosophical Manuscripts of 1844. Collected Works. Volume III*, Lawrence & Wishart, London.
Marx, L. (1964) *The Machine in the Garden. Technology and the Pastoral Ideal in America*, Oxford University Press, New York.
Massumi, B. (1988) Translator's foreword to G. Deleuze and F. Guattari *A Thousand Plateaus. Capitalism and Schizophrenia*, Atlhone Press, London, pp. i–xv.
Massumi, B. (1992a) *A User's Guide to Capitalism and Schizophrenia. Deviations from Deleuze and Guattari*, MIT Press, Cambridge, Mass.

Massumi, B. (1992b) Everywhere you want to be. Introduction to fear, in J. Broadhurst (ed.) *Deleuze and the Transcendental Unconscious*, University of Warwick Press, Coventry, pp. 175–216.

Matless, D. (1991) A modern stream: an essay in water, landscape, modernism and geography, *Environment and Planning D. Society and Space*, Vol. 9, pp. 280–93.

McCafferty, L. (ed.) (1991) *Storming the Reality Studio*, Duke University Press, Durham, N.C.

Melbin, M. (1987) *Night as Frontier. Colonizing the World after Dark*, Free Press, New York.

Michaux, H. (1992) *Spaced, Displaced*, Bloodaxe Books, Newcastle upon Tyne.

Mills, S. (1991) *Discourses of Difference. An Analysis of Women's Travel Writing and Colonialism*, Routledge, London.

Minh-ha, Trinh, T. (1989) *Women, Nature, Other. Writing, Postcoloniality and Feminisim*, Indiana University Press, Bloomington.

Mishra, V. and White, B. (1991) What is post(-)colonialism? *Textual Practice*, Vol. 5, pp. 399–414.

Montrose, L. (1991) The work of gender in the discourse of discovery, *Representations*, Vol. 33, pp. 1–36.

Morley, D. (1991) Where the global meets the local: notes from the sitting room, *Screen*, Vol. 32, pp. 1–15.

Morris, M. (1987) At Henry Parkes Motel, *Cultural Studies*, Vol. 3, pp. 1–36.

Morris, M. (1992) On the beach, in L. Grossberg, C. Nelson and P. Treichler (eds.) *Cultural Studies*, Routledge, New York, pp. 450–78.

Mulgan, G. J. (1991) *Communication and Control. Networks and the New Economics of Communication*, Guilford Press, New York.

Mumford, L. (1966) *Technics and Civilisation.*

Musselwhite, M. (1987) *Partings Welded Together. Politics and Desire in the Nineteenth Century English Novel*, Methuen, London.

Nasaw, D. (1992) Cities of light, landscapes of pleasure, in D. Ward and O. Zunz (eds.) *The Landscape of Modernity*, Russell Sage Press, New York, pp. 273–86.

Naylor, G. (1992) Design, craft and identity, in B. Ford (ed.) *Victorian Britain. Volume 7. The Cambridge Cultural History*, Cambridge University Press.

Nietzsche, F. (1967) *Werke*, Vol. 4, no. 2.

Nye, D. (1990) *Electrifying America. Social Meanings of New Technology*, MIT Press, Cambridge, Mass.

O'Riordan, T. (1992) Review, *The Higher*, 26 June, p. 16.

Ousby, I. (1991) *The Englishman's England: Travel, Taste, and the Rise of Tourism*, Cambridge University Press.

Parkes, D. N. and Thrift, N. J. (1979) *Time, Spaces, Places. A Chronogeographic Perspective*, Wiley, Chichester.

Penley, C. and Ross, A. (eds.) (1991) *Technoculture*, University of Minnesota Press, Minneapolis.

Perez, R. (1990) *On Anarchy and Schizoanalysis*, Automedia, New York.

Pickering, A. (ed.) (1993) *Science as Practice and Culture*, University of Chicago Press.

Piercy, M. (1991) *Body of Glass* (in the USA, *He, She and It*), Knopf, New York.
Plant, S. (1992) *The Most Radical Gesture. The Situationist International in a Postmodern Age*, Routledge, London.
Plowden, W. (1971) *The Motor Car and Politics in Britain*, Penguin, Harmondsworth.
Pomeroy, J. (1991) Black box s-thetix: labour, research and survival in the he(art) of the beast, in C. Penley and A. Ross (eds.) *Technoculture*, University of Minnesota Press, Minneapolis.
Porter, D. (1991a) *Haunted Journeys. Desire and Transgression in European Travel Writing*, Princeton University Press.
Porter, R. (1991b) The history of photography, in P. Burke (ed.) *New Approaches to Social History*, Methuen, London.
Poster, M. (1990) *The Mode of Information. Poststructuralism and Social Context*, Polity Press, Cambridge.
Pratt, G. B. (1992) Spatial Metaphors and Speaking Positions, *Environment and Planning D. Society and Space*, no. 10, p. 241–4.
Pratt, M. L. (1991) *Imperial Eyes. Travel Writing and Transculturation*, Routledge, London.
Prendergast, C. (1992) *Paris and the Nineteenth Century*, Blackwell, Oxford.
Preston, P. (1992) Artificial Life *Times Higher Education Supplement*, March 16.
Price, P. (1990) *Alabi's World*. Harvard University Press, Cambridge, Mass.
Probyn, E. (1993) *Sexing the Self. Gendered Positions in Cultural Studies*, Routledge, London.
Radway, J. (1988) Reception study: ethnography and the problems of dispersed audiences and nomadic subjects, *Cultural Studies*, Vol. 2, pp. 359–76.
Rigby, B. (1991) *Popular Culture in Modern France. A Study of Cultural Discourse*, Routledge, London.
Robins, K. F. (1991) Prisoners of the city: whatever could a postmodern city be? *New Formations*, no. 15, pp. 1–22.
Robson, B. T. (1973) *Urban Systems*, Methuen, London.
Robson, B. T. (1990) The Years Between, in R. Dodgshon, and R. A. Butlin, (eds.) *A New Historical Geography of England and Wales*, Academic Press, London, pp. 545–78.
Rojek, C. (1993) *Ways of Escape. Modern Transformations in Leisure and Travel*, Macmillan, London.
Ronell, A. (1989) *The Telephone Book*, University of Nebraska Press, Nebraska.
Rosaldo, R. (1988) *Culture and Truth. The Remaking of Social Analysis*, Beacon Press, Boston.
Rose, G. (1993) *Feminism and Geography*, Polity Press, Cambridge.
Ross, A. (1991) *Strange Weather. Culture, Science and Technology in the Age of Limits*, Verso, London.
Said, E. W. (1978) *Orientalism*, Penguin, Harmondsworth.
Said, E. W. (1984) Travelling theory, in *The World, the Text and the Critic*, Faber & Faber, London.

Sauer, C.O. (1956) The education of a geographer, *Annals of the Association of American Geographers*, Vol. 46, pp. 287–99.
Sayer, A. (1989) The new regional geography and the problem of narrative, *Environment and Planning D. Society and Space*, Vol. 7, pp. 253–76.
Schivelsbuch, W. (1986) *The Railway Journey. The Industrialisation of Time and Space in the 19th Century*, University of California Press, Berkeley.
Schivelsbuch, W. (1988) *Disenchanted Night. The Industrialisation of Light in the Nineteenth Century*, University of California Press, Berkeley.
Scott, J. W. (1992) Experience, in J. Butler and J. W. Scott (eds.) *Feminists Theorise the Political*, Routledge, London, pp. 22–40.
Seltzer, M. (1992) *Bodies and Machines*, Routledge, New York.
Shapiro, M. J. (1992) *Reading the Postmodern Polity. Political Theory as Textual Practice*, Minnesota University Press, Minneapolis.
Shilling, C. (1993) *The Body and Social Theory*, Sage, London.
Silverstone, R. and Hirsch, M. (eds.). (1992) *Consuming Technologies*, Routledge, London.
Simmel, G. (1978) *The Philosophy of Money*, Routledge & Kegan Paul, London.
Sklair, L. (1991) *Global Sociology*, Macmillan, London.
Slater, D. (1992) On the borders of social theory *Environment and Planning D. Society and Space*, Vol. 10, pp. 307–28.
Slusser, G. and Rabkin, E. S. (eds.) (1992) *Styles of Creation. Aesthetic Technique and the Creation of Fictional Worlds*, Georgia University Press, Athens.
Slusser, G. and Shippey, T. (eds.) (1992) *Fiction 2000. Cyberpunk and the Fiction of Narrative*, Georgia University Press, Athens.
Spivak, G. C. (1987) *In Other Worlds. Essays in Cultural Politics*, Methuen, London.
Spivak, G. C. (1990) *The Post-Colonial Critic. Interviews, Strategies, Dialogues*, Routledge, New York.
Strathern, M. (1992a) *After Nature. English Kinship in the Late Twentieth Century*, Cambridge University Press.
Strathern, M. (1992b) Writing societies, writing persons, *History of the Human Sciences*, Vol. 5, pp. 5–16.
Strathern, M. (1992c) Reproducing anthropology in S. Wallman (ed.) *Contemporary Futures*, Routledge, London, pp. 172–89.
Synott, A. (1993) *The Body Social*, Routledge, London.
Taussig, M. (1993) *Mimesis and Alterity. A Particular History of the Senses*, Routledge, New York.
Thrift, N. J. (1990) The geography of capitalist time consciousness, in J. Hassard (ed.) *The Sociology of Time*, Macmillan, London.
Tichi, C. (1987) *Shifting Gears. Technology, Literature, Culture in Modernist America*, University of North Carolina Press, Chapel Hill.
Tomas, D. (1992) From gesture to activity: dislocating the anthropological scriptorium, *Cultural Studies*, Vol. 6, pp. 1–26.
Urry, J. (1990) *The Tourist Gaze*, Sage, London.
van den Abbeele, G. (1992) *Travel as Metaphor. From Montaigne to Rousseau*, University of Minnesota Press, Minneapolis.

Vasseleu, C. (1991) Life itself, in R. Diprose and R. Ferrell (eds.) *Cartographies*, Allen & Unwin, Sydney.
Virilio, P. (1975) *Cause Commune, Nomades et Vagabonds*, UGE, Paris.
Virilio, P. (1977) *Vitesse et Politique*, Galilée, Paris.
Virilio, P. (1980) *Esthetique de la Disparition*, Ballard, Paris.
Virilio, P. (1986) *Speed and Politics*, Semiotext(e), New York.
Virilio, P. (1988) *La Machine de Vision*, Galilée, Paris.
Virilio, P. (1989) *War and Cinema. The Logistics of Perception*, Verso, London.
Virilio, P. (1990) *L'Inertie Polaire*, Galilée, Paris.
Virilio, P. (1991a) *The Aesthetics of Disappearance*, Semiotext(e), New York.
Virilio, P. (1991b) *The Lost Dimension*, Semiotext(e), New York.
Wagner, R. (1991) The fractal person, in M. Godelier and M. Strathern (eds.) *Big Men and Great Men. Personifications of Power in Melanesia*, Cambridge University Press.
Ward, D. and Zunz, O. (eds.) (1992) *The Landscape of Modernity*, Russell Sage Press, New York.
Wilcox, D. J. (1987) *The Measure of Times Past. Pre-Newtonian Chronologies and the Rhetoric of Relative Time*, Chicago University Press.
Williams, R. (1954) *Preface to Film*, Film Drama, London.
Williams, R. (1977) *Marxism and Literature*, Oxford University Press, London.
Williams, R. (1979) *Politics and Letters*, Verso, London.
Williams, R. (1981) *Culture*, Fontana, London.
Williams, R. (1990) *Notes on the Underground. An Essay on Technology, Society and the Imagination*, MIT Press, Cambridge, Mass.
Wilson, E. (1992) *The Sphinx in the City*, Virago, London.
Wollen, C. (1993) *Raiding the Ice Box. Reflections on Twentieth Century Culture*, Verso, London.
Wood, D. (1990) *Philosophy at the Limit*, Unwin Hyman, London.
Woolf, J. (1993) On the road again: metaphors of travel in cultural criticism, *Cultural Studies*, Vol. 7, pp. 224–39.
Young, R. (1990) *White Mythologies. Writing History and the West*, Routledge, London.

Index